ENERGY SCIENCE, ENGINEERING AND TECHNOLOGY

A COMPREHENSIVE GUIDE TO ENERGY PRODUCTION AND DEVELOPMENT

ENERGY SCIENCE, ENGINEERING AND TECHNOLOGY

Additional books and e-books in this series can be found on Nova's website under the Series tab.

Energy Science, Engineering and Technology

A Comprehensive Guide to Energy Production and Development

Shannon Alvarado
Editor

Copyright © 2020 by Nova Science Publishers, Inc.

All rights reserved. No part of this book may be reproduced, stored in a retrieval system or transmitted in any form or by any means: electronic, electrostatic, magnetic, tape, mechanical photocopying, recording or otherwise without the written permission of the Publisher.

We have partnered with Copyright Clearance Center to make it easy for you to obtain permissions to reuse content from this publication. Simply navigate to this publication's page on Nova's website and locate the "Get Permission" button below the title description. This button is linked directly to the title's permission page on copyright.com. Alternatively, you can visit copyright.com and search by title, ISBN, or ISSN.

For further questions about using the service on copyright.com, please contact:
Copyright Clearance Center
Phone: +1-(978) 750-8400   Fax: +1-(978) 750-4470   E-mail: info@copyright.com

NOTICE TO THE READER

The Publisher has taken reasonable care in the preparation of this book, but makes no expressed or implied warranty of any kind and assumes no responsibility for any errors or omissions. No liability is assumed for incidental or consequential damages in connection with or arising out of information contained in this book. The Publisher shall not be liable for any special, consequential, or exemplary damages resulting, in whole or in part, from the readers' use of, or reliance upon, this material. Any parts of this book based on government reports are so indicated and copyright is claimed for those parts to the extent applicable to compilations of such works.

Independent verification should be sought for any data, advice or recommendations contained in this book. In addition, no responsibility is assumed by the Publisher for any injury and/or damage to persons or property arising from any methods, products, instructions, ideas or otherwise contained in this publication.

This publication is designed to provide accurate and authoritative information with regard to the subject matter covered herein. It is sold with the clear understanding that the Publisher is not engaged in rendering legal or any other professional services. If legal or any other expert assistance is required, the services of a competent person should be sought. FROM A DECLARATION OF PARTICIPANTS JOINTLY ADOPTED BY A COMMITTEE OF THE AMERICAN BAR ASSOCIATION AND A COMMITTEE OF PUBLISHERS.

Additional color graphics may be available in the e-book version of this book.

Library of Congress Cataloging-in-Publication Data

Names: Alvarado, Shannon, editor.
Title: A Comprehensive Guide to Energy Production and Development
Description: New York: Nova Science Publishers, [2019] | Series: Energy Science, Engineering and Technology | Includes bibliographical references and index.
Identifiers: LCCN 2019954667 (print) | ISBN 9781536167283 (paperback) |
  ISBN 9781536167498 (adobe pdf)

Published by Nova Science Publishers, Inc. † New York

CONTENTS

Preface		vii
Chapter 1	Guidelines for Promotion of Renewable Energy in Terms of Knowledge-Based Economy *Valentinas Klevas, Lina Murauskaite and Audrone Kleviene*	1
Chapter 2	Transformation Pathways Towards a Clean, Secure and Efficient European Energy System: A MCDA Approach *Charikleia Karakosta and Aikaterini Papapostolou*	35
Chapter 3	Electricity Transitions in Germany: The Transformation of a Strategic Action Field *Gerhard Fuchs*	51
Chapter 4	Techno-Economic Analysis of Standalone Hybrid Energy System *Shweta Goyal, Sachin Mishra, and Anamika Bhatia*	75

Chapter 5	Design and Manufacturing Approaches to Improve Reliability of Wind Turbine Gearboxes: A Review *Wei Jiang*	**165**
Index		**193**
Related Nova Publications		**199**

PREFACE

In A Comprehensive Guide to Energy Production and Development, a comprehensive analysis of existing examples in international practice regarding methods for massive implementation of renewable energy technology is provided.

Following this, the authors assess alternative pathways for a future European energy system through the active involvement of relevant stakeholders according to their performance in key areas, such as the regulatory framework, market maturity, economic factors and stakeholder awareness towards a sustainable energy system, by using a multicriteria decision aid method.

Recent developments in the field of electricity generation and distribution in Germany are presented, and decentralized forms of electricity generation and distribution are analyzed.

As the consumption of power is increasing day by day, the power generation requirement is also increasing. As such, the authors discuss how renewable systems may be very useful in remote areas.

The concluding paper studies the failure modes of wind turbine gearboxes and their causes. Factors affecting the gearbox reliability in both design and manufacturing processes are then analyzed.

Chapter 1 - Nowadays many facts show emerging environmental threats mainly due to emissions of fossil fuel combustion. The problem of climate change became more acute following historic adoption of the Paris Agreement by 195 countries in December 2015, which is dealing with the mitigation of greenhouse gases emissions. United Nations announced that 185 countries have ratified Paris Agreement before March 2019; therefore, revolutionary measures and assumptions are needed. It is generally recognized that renewable energy sources (RES) technologies are the main measures for environmental protection, energy supply for future generations and contribute to the solutions of social problems. Although the development of RES technologies is impressive, eventually new problems arise, which are caused by insufficiently harmonized knowledge about not only positive but also negative consequences of the use of RES. This is particularly the case with disproportionate use of RES because biomass dominates by other types of RES, especially in the heating sector. A theoretical breakthrough of the knowledge-based economy towards the assessment of regional economic progress allows highlighting the main methodological direction, which focuses on harmonious RES technology integration replacing fossil fuels in the heating sector. Nowadays a growing number of experts recognize the importance of RES integration into local energy networks that could be best implemented on the local level. District heating (DH) infrastructure is recognized as one of the most favorable for integrating various RES technologies. Improvement of the infrastructure in parallel with the use of RES technologies has critical importance. The infrastructure of district heating systems is essential for the heating sector towards the wider use of different types of RES.

Comprehensive analysis of existing examples in international practice regarding methods for massive implementation of RES technologies was provided. The main accent was the methods of improvement of the infrastructure that is necessary for the wide use of RES technologies.

The main outcome of the research is the elaboration of guidelines for the implementation of strategic energy goals in the heating sector by improvement of the infrastructure, which is favorable for the integration of all types of RES technologies, thus increasing the security of energy supply, stabilizing energy prices, reducing air pollution in cities, etc.

Chapter 2 - The European Union's (EU) energy, innovation and climate challenges define the direction of a future European energy system, but the specific technology pathways are policy sensitive and require careful comparative evaluation. The Strategic Energy Technology Plan (SET-Plan), as part of the technology pillar of European Energy and Climate Policy, has identified strategic energy technologies and designed roadmaps to tackle the individual barriers hindering their effective implementation across Europe. In the above context, a set of alternative pathways were designed by identifying factors that conform to the objectives of the EU and the SET-plan, in particular competitiveness and a clean, secure and efficient energy system. These pathways are positioned under two key uncertainties; the level of cooperation (i.e., cooperation versus entrenchment) and the level of decentralisation (i.e., decentralisation versus path dependency). The aim of this paper is to assess these pathways, through the active involvement of relevant stakeholders, according to their performance in key areas, such as the regulatory framework, market maturity, economic factors and stakeholder awareness towards a sustainable energy system, by using a multicriteria decision aid method (MCDA). Key challenge of the authors' research is to assist European policy makers in drawing recommendations by exploring important elements, drivers and factors of the energy transition and the long term impacts of alternative mitigation options on the economy, the energy sector and technology development. The results could prove to be useful in further supporting strategic decision making in Europe's energy sector towards a clean, secure and efficient energy system.

Chapter 3 - The chapter draws on recent developments in the field of electricity generation and distribution in Germany. The authors analyze decentralized forms of electricity generation and distribution. Pioneers of this development seized opportunities connected with broad institutional changes to discredit the status quo and work out legitimations for their new model of how to generate and distribute electricity. The authors' analysis suggests important differences in how actors legitimate novel organizational forms in emerging vs. mature fields and it underscores the need for theories of social change that explicitly account for field context.

Chapter 4 - Consumption of power is increasing day by day so the more power generation requirement is necessary. This generation cannot be handling by non-renewable sources individually; because they will deplete after certain period of time. So the some gap between power supply and demand can be cover-up by the renewable sources. Renewable system is very useful for off-grid system for remote area to build without having complicated grid system.

Rural electrification through main grid increases investment cost and losses. This problem can be overcome by renewable standalone system. The problem with the renewable sources is that they not able to give continue energy so hybrid energy system can be used to decentralize the problem. The term hybrid system describes by the combination of two or more renewable sources with fossil fuel power diesel/petrol to provide continue energy. Hybrid energy system is very useful in location where extension of electrical grid is not possible and it may say very difficult.

Hybrid energy system has found much wider distribution than just as individual stand-alone renewable system for electrification of rural area. Hybrid system is very useful for remote area. But the planning of rural electrification is very difficult; it needs very deep economic, technical and social study of the location. It is very important to develop a proper simulation model and optimization techniques. Wide range of different configuration is possible but choice of modeled configuration must suit the location. It is necessary to have the knowledge about energy demand and resources available for specific location MATLAB script is developed which calculates the optimum system design and allows the user to evaluate the electro-economic and technical feasibility of a large number of technologies. The basic principle is to minimize the total cost of energy (COE) of the system while satisfying the unit and system constraints such as number of solar modules, diameter of parabolic reflectors, number of solar collectors, gross area available hub height of wind turbine, rotor diameter of wind blades, scale factor, area of the installation.

There are a huge number of MPPT techniques and algorithms have come in to existence as discussed in the literature. Most commonly employed technique among them is P&O algorithm due to its simplicity and robust nature. Proposed Artificial BEE colony (ABC) inspired MPPT algorithm has therefore been compared to P&O algorithm. For this new Swarm intelligence based Artificial Bee Colony (ABC) algorithm has been used to track MPP of PV module. Objective function is based on voltage power characteristics of PV module. The constraints taken under consideration include power of the PV module and battery. ABC and P&O aim at tracking maximum power output and minimizing the objective function to global optimal result. Algorithm traces global maxima and minima by voltage versus power curve. Power balance constraints for any period includes power supplied by the hybrid generation and load demand, i.e., generated power must meet total load demand, satisfying power quality and reliability constraints.

Comparative curve for MPPT techniques using ABC and P&O using power curve for PV irradiance shows that ABC algorithm has better power yield as compared to P&O. P&O is the most robust and simple in application but lacks exploitation of PV power at higher irradiance. At normal irradiance below 650 W/m^2, P&O has been found to exhibit better results while for higher irradiance, ABC gives higher power output.

Generation unit and storage unit capacity depends on load demand. Load data is the actual load as obtained from energy meter reading for the months of the year 2014. Hourly load pattern has been observed for typical days and hourly load profile has been fabricated accordingly. Daily base load was estimated to be 132 kWh/day and peal load of around 35 kW with day-to-day random variability of 15, time step random variability of 20% and load factor of 0.3. Annual average load found to be 132 and peak load as 28.9 kW in the month of August. The obtained hourly load profile during months of year 2014.

No fossil fuel based power source has been considered in this study and thus renewable fraction is .07% but when no fossil fuel used renewable fraction is 100%.The COE found in optimization result is Rs.45.46 and Rs 19.45 in case study 1 (with fossil fuel) and case study 2 (with fossil fuel) respectively, which is less than present COE (grid connection and diesel generator) of Rs. 20.20 and likely to increase in near future as per decreasing COE trend of RES based generation. Obtained COE is competitive COE with hybrid PV-wind-bio-diesel system Rs. 19.45/kWh, hybrid PV-Wind-Diesel Rs. 13.46/kWh, hybrid PV-bio-diesel Rs. 17.71/kWh, hybrid wind-diesel Rs. 14.21/kWh and hybrid wind –bio-diesel Rs. 33.23 /kWh.

Chapter 5 - Renewable energy has gained increasing interest worldwide against energy crisis and environmental pollution. As a renewable and clean source of energy, wind energy power generation has had an exponential growth in the last decade. However, for a typical design life of 20 years of wind turbines, the premature failure of wind turbine gearboxes greatly increase the downtime, maintenance expenditures, and cost of energy.

Wind turbine gearboxes are one of the most expensive components in wind turbine system. Operational experiences reveal that gearboxes are the weakest link in multi-megawatt wind turbines. This paper first studies the failure modes of wind turbine gearboxes and their causes. Factors affecting the gearbox reliability in both design and manufacturing processes are then analyzed. Measures to improve the reliability and prolong the service life of MW sized wind turbine gearboxes through innovative design and manufacturing approaches are finally proposed. Hopefully, by advanced design and manufacturing, high reliability gearboxes can be developed to operate confidently under complex loading conditions.

In: A Comprehensive Guide to Energy ... ISBN: 978-1-53616-728-3
Editor: Shannon Alvarado © 2020 Nova Science Publishers, Inc.

Chapter 1

GUIDELINES FOR PROMOTION OF RENEWABLE ENERGY IN TERMS OF KNOWLEDGE-BASED ECONOMY

Valentinas Klevas[*], *Lina Murauskaite*
and Audrone Kleviene
Lithuanian Energy Institute, Kaunas, Lithuania

ABSTRACT

Nowadays many facts show emerging environmental threats mainly due to emissions of fossil fuel combustion. The problem of climate change became more acute following historic adoption of the Paris Agreement by 195 countries in December 2015, which is dealing with the mitigation of greenhouse gases emissions. United Nations announced that 185 countries have ratified Paris Agreement before March 2019; therefore, revolutionary measures and assumptions are needed. It is generally recognized that renewable energy sources (RES) technologies are the main measures for environmental protection, energy supply for future generations and contribute to the solutions of social problems. Although the development

---

[*] Corresponding Author's E-mail: Valentinas.Klevas@lei.lt.

of RES technologies is impressive, eventually new problems arise, which are caused by insufficiently harmonized knowledge about not only positive but also negative consequences of the use of RES. This is particularly the case with disproportionate use of RES because biomass dominates by other types of RES, especially in the heating sector. A theoretical breakthrough of the knowledge-based economy towards the assessment of regional economic progress allows highlighting the main methodological direction, which focuses on harmonious RES technology integration replacing fossil fuels in the heating sector. Nowadays a growing number of experts recognize the importance of RES integration into local energy networks that could be best implemented on the local level. District heating (DH) infrastructure is recognized as one of the most favorable for integrating various RES technologies. Improvement of the infrastructure in parallel with the use of RES technologies has critical importance. The infrastructure of district heating systems is essential for the heating sector towards the wider use of different types of RES.

Comprehensive analysis of existing examples in international practice regarding methods for massive implementation of RES technologies was provided. The main accent was the methods of improvement of the infrastructure that is necessary for the wide use of RES technologies.

The main outcome of the research is the elaboration of guidelines for the implementation of strategic energy goals in the heating sector by improvement of the infrastructure, which is favorable for the integration of all types of RES technologies, thus increasing the security of energy supply, stabilizing energy prices, reducing air pollution in cities, etc.

Keywords: renewable energy sources; knowledge-based economy; transformation of energy infrastructure of municipalities

INTRODUCTION

Progress of the knowledge-based economy theory is expressed by the significant impact in various aspects. Numerous studies have determined the main cause of climate change such as the emissions from fossil fuel combustion, especially CO_2. This demonstrates the importance of knowledge for proving the facts which are often controversial. The historic Paris Agreement was achieved due to detailed and proven knowledge on the effects of fossil fuel combustion when comprehensive knowledge was properly presented. Therefore, the accumulation of scientific knowledge about the impact of fossil fuel combustion emissions on the environment has allowed convincing politicians about necessary rescue measures globally. Knowledge gained from the research and their synthesis has enabled highlighting the sources of ecological threats: "Now we have enough data indicating that such critical environmental problems as a loss of biodiversity, land degradation, climate change, and biosphere pollution strongly correlate with the growth of knowledge" [1]. On the other hand, the main direction has emerged, which may have a critical impact on escaping from climate change: it is the replacement of fossil fuels with renewable energy sources (further in the text RES). "As a result of the increased awareness of the dangers posed by global climate changes (mainly caused by growing global energy consumption needs), the quest for clean and sustainable energy future is becoming of paramount importance. This can be largely realized via a large-scale integration of variable RES, such as wind and solar, which have relatively low-carbon footprints" [2]. In recent years, consumption of fossil fuel and mitigation of climate change have become major challenges for governments all over the world. To engage these challenges, many countries are pursuing research and development of RES technologies [3]. In the past few years, the use of RES rapidly increased all over the world. RES have been recognized as significant alternative energy sources for the implementation of energy diversification. Political support for RES has been growing continuously both at the national and international levels. The industry of renewables could be important to generate employment and stimulate growth [4]. Investment in RES may bring considerable profits; therefore, more and more enterprises will be involved in this field. The successful commercialization of indigenous, non-fossil energy resources is expected to promote regional economic development and employment, enable the increase of national energy security and reduce a significant part of the increasing trade deficit for the import of fossil fuels [5].

However, a new stimulus for the successful uptake of RES technologies is needed. The main aspects of this stimulus might be multidisciplinary knowledge of the positive impact of RES and dissemination of it to all interested government and population groups.

"Renewable energy technologies are on the verge of a new era. In many countries and regions, renewable energy is already responsible for meeting a substantial share of energy demand.

Progress in building new energy systems is already considerable. But issues like energy independence, eradicating energy poverty, combating climate change and improving the crisis-robustness of energy systems are asking to accelerate the deployment of renewables" [6].

The development of the knowledge-based economy has given undeniable impetus to sustainable development movement, which was started and stimulated by the problems of energy development. In 1987, the World Commission on Environment and Development led by Norwegian Prime Minister Gro Harlem Brundtland united environmental and economic development perspectives into the concept of sustainable development, which quickly gained international recognition as an advanced human-economic development strategy for the millennium. Its content highlights four main areas of energy development: a) the development of energy without affecting the balance of nature; b) the energy self-sufficiency for future generations; c) the reliability of energy supply in a broad sense and security; d) from a social point of view, it is the opportunity to provide energy for all segments of the population regardless of their place of residence (city, village). Development of RES actually covers all aspects of sustainable development: security of energy supply, environmental protection, solution of social issues, and energy self-sufficiency for the future generations.

Economic research of RES has increased recently through the implementation of European Union directives, with a particular emphasis on the role of territorial aspect (cities, districts). Promotion for widespread use of RES is declared both in the European Union and in Lithuanian laws. Adoption of the Directive 2009/28/EC "On the promotion of the use of energy from renewable sources" [7] required the development of a National Renewable Energy Action Plan for 2010-2020 and many other documents, including the local level. This indicates that the development of RES takes an increasingly significant role in the energy policy of all Member States, including Lithuania.

However, the practical application of the conceptual statements of the knowledge-based economy significantly lags. Many authors almost unanimously stated that this lag could be reduced only if research and practical implementation will be linked to the progress of specific regions and cities. Therefore, one of the most important modern directions of the development of theoretical knowledge of the knowledge-based economy is the highlight of the regional territorial context. "This growing emphasis upon knowledge has been influential in shaping new conceptions of urban and regional development policy and strong claims have been made as to the potential of such developments for urban and regional regeneration, linking the knowledge-based competitive advantage of firms with that of territories defined at various spatial levels" [8].

Consequently, the concept of promotion of RES in terms of knowledge-based economy in practice means that the process of the use of RES technologies must be controlled and adjusted on all relevant aspects in the long term. The incentive system is determined by the entire chain of energy supply and consumption and is depending on the specific benefit to be assessed that emphasizes the right distribution of cost and income.

However, there are still problems which have to be analyzed and evaluated; also rational solutions should be provided. "In many power systems, the level of integration of such resources is dramatically increasing. However, their intermittent nature poses significant challenges in the predominantly conventional power systems that currently exist. Generally, the higher the integration level of intermittent power sources is, the higher the flexibility needs are in the system under consideration." [2]. This means that development of the infrastructure in parallel with the use of RES technologies has critical importance. Moreover, the infrastructure of district heating (further in the text DH) systems is very important for the heating sector in order to widely use the different types of RES.

Municipalities could play an important role in promoting assumptions for sustainable energy development because the local government performs a number of functions in the energy sector, and these opportunities should be used. On the other hand, structures that could concentrate various financial resources, which are available on the state level, are scarce in cities.

These financial resources together with private investment would allow the development of renewable energy technologies on the significantly larger scale.

METHODOLOGY FOR RENEWABLE ENERGY SOURCES IMPLEMENTATION IN LARGE SCALE

Introductory Remarks

The main issues of economic research and problems are as follows:

- Currently, RES are considered as an integral entity, and all advantages and disadvantages are generalized, but at the same time there are missing links with specific conditions and available infrastructure. Therefore, one of the most relevant issues is dominance of one RES type in Lithuania under present economic conditions. The main focus is concentrated for the use of biomass, which reasons are not properly investigated and determined. Meanwhile, alternative RES types, such as solar, wind or geothermal energy, are relatively rarely used.
- In this context, cities and towns with their DH systems gain vital significance now and for the future. At present, it is almost universally recognized that DH infrastructure is one of the most favorable means of integrating various RES technologies in the heat sector. Therefore, it is very important to analyze the assumptions (legal, economic and technical) that will allow the integration of RES technologies to a large extent by using DH infrastructure.
- Formation and presentation of knowledge structure concerning the public benefit of the use of RES technologies and dissemination of this knowledge has particular importance for massive implementation of RES technologies [9-13].

Identifying the Problem of Heating Sector: Lithuanian Case

Practical examples in Europe and Asia show the rapid and substantial progress of renewable energy in recent years. The development has been driven by policies of local, national and regional authorities, in close cooperation with the business community, as well as continued technological innovation and cost reduction in energy generated by RES technologies.

However, there is a particularly important problem in the heat sector: the chosen pathway of biomass aims at a short-term effect, which in this case means one-sided use of RES.

"So here is nowadays' challenge for policymakers and decision makers: how to pass the threshold in the short term in order to prepare for the longer term? Achieving energy systems that will meet tomorrow's energy demand in a sustainable, responsible way is possible, as some countries already prove. But further deployment requires large effort on the part of policymakers and business leaders" [6].

The main issue that must be solved is the one-sided use of RES in Lithuanian heat sector, where the focus is set mainly to biomass. Unfortunately, other types of RES, such as solar or geothermal energy, are scarcely used.

Figure 1 shows the state of the use of RES technologies in Lithuania.

This is a consequence of unilateral and primitive approach to RES by assuming that biomass is identical to solar or geothermal energy. Such controversial phenomena occur especially in the heating sector, where disproportion of the use of RES has deepened. Currently, undue prominence is given for the future projections of biomass, which is the energy source from the outcome of economic activity. Increase of biomass is the result of continuing economic activity. The potential of inexhaustible solar and geothermal energy is undervalued at present decisions in Lithuania.

The data in Table 1 show the percentage of the use of RES technologies by regions in the heating sector. Biomass has a dominant position in all Lithuanian regions because the combustion of biomass is treated as neutral in terms of CO_2 emissions. However, this is a serious challenge to the sustainable development of energy and a threat to forests and their ecosystems.

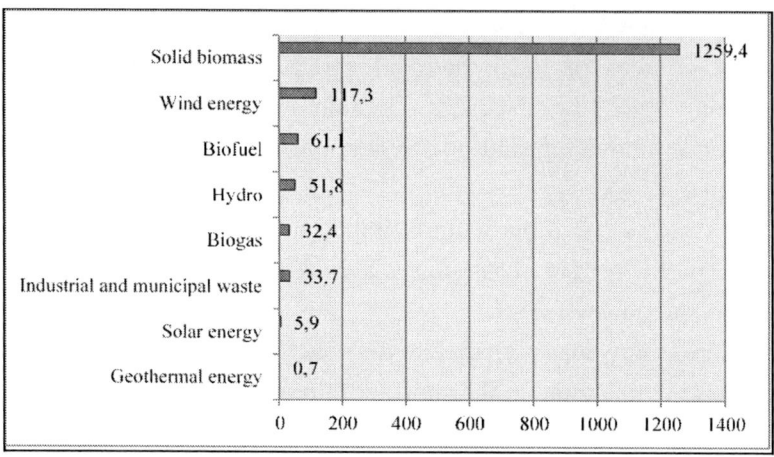

Figure 1. RES consumption in Lithuania in 2017, ktoe.

Table 1. Disproportions of the use of RES technology by Lithuanian regions in the heating sector (installed capacity, kW)

	Solar energy	Geothermal energy	Straw	Biogas	Biomass	Landfill gas	Municipal waste
Region Vilnius			0,3		284,7		
Region Kaunas	7,1		2,9	5,9	359,1	1,6	
Region Klaipeda		0,1		0,9	220,2		50
Region Siauliai			1,2	0,8	134,2		
Region Panevezys				5	149,9		
Region Alytus					96,7		
Region Telsiai					111,7		
Region Utena					114,6		
Region Marijampole				0,9	73,6		
Region Taurage					70,1		

The research issue is the assessment of RES social utility, which could demonstrate benefits that are underestimated in the investment decisions. Significant advantages of RES are inexhaustibility and sufficiency of energy resources for the future generations. Moreover, the use of such RES technologies as solar energy helps in solving environmental issues. Therefore, additional finances could be allocated for RES development. Variety of the use of RES opportunities and incentives remains the key aspect because the sufficient supply of energy for the future generations is guaranteed by the use of RES.

Solving Problems Related to the Use of RES in Large Scale Using DH Infrastructure: A Local Approach

An important aspect to be considered in this analysis is that the achievement of sustainable development goals is related to the territorial aspect by many authors. Agenda 21 Action Plan and the Rio Declaration on Environment and Development originally presented the concept of sustainable development for nations at the 1992 Rio Summit. Nowadays a growing number of experts recognize the importance of RES integration into local energy networks that could be best implemented on the local level, such as municipalities or cities [14-15]. It is universally recognized that DH infrastructure is one of the most favorable means for the integration of various RES technologies.

The methodological principle is expressed by this Hudson quote: "The future success of these territories is seen to depend on the continuing production of new knowledge, translating this into innovative products and processes and maintaining first-mover advantages. The task of public policy is then to try to ensure that the knowledge-based process of moving forward, ever onward and upward, is facilitated in successful places and that the lessons of their success are translated to unsuccessful places, in so far as this is possible within the parameters of capitalist social relations" [8].

There are plenty of examples in worldwide practices which show that disproportionality can be avoided by purposeful measures of the state policy. Good examples are North European countries: Denmark and Sweden.

Therefore, it could be concluded that the problems of the heating sector remain and the main directions of the solutions are a) analysis of DH systems as a potential infrastructure for the diversification of RES technologies, b) analysis of knowledge dissemination tools and formation of RES programs model for cities.

However, there are very few empirical economics papers written on DH, as noted by [16]. Some authors have analyzed sustainable development aspects of the use of RES technologies [17, 18]; and promotion of RES in Lithuania [5, 19]. However, DH is usually analyzed in the engineering field of scientific papers related to combined heat and power plants [20-22] or policy aspects of DH [5, 23, 24] in Lithuanian and foreign authors research.

Evolution of energy and industry has considerably affected the operation of DH systems (decline of total heat consumption, disconnection of industries from DH systems, disconnection of individual consumers, development of new customers, implementation of automated consumer substations, etc.). A retrospective analysis of the evolution of the heating sector in Lithuania is examined by [25].

However, the progress of DH systems that takes place in foreign countries, especially in Northern Europe (Denmark, Sweden, and Finland) shows the increasing integration of RES technologies in DH systems. International scientific papers, mainly from Scandinavian countries, in recent years aims at analyzing 4th generation DH [26], low-temperature DH [27], and smart DH network [26, 28, 29].

The main economic and practical reason which restricts the use of RES in new EU Member States is ignoring the DH as natural monopoly status. It should be noted that DH companies on a natural monopoly aspect have been scarcely analyzed in Lithuania, although it is one of the most important reasons why such a favorable infrastructure has a low level of utilization of RES technologies. In Lithuanian heating sector, the use of RES is exceptionally limited to biomass, which only partially could be treated as a renewable resource. A natural monopoly is a unique economic condition which could not be equalized to a free market economy, especially for DH infrastructure. The natural monopoly is determined by regulation, pricing tools and methods. DH overcame reforms in post-Soviet legacy with major changes of fuel prices and conversion to profit-making organizations despite the natural monopoly status. Authors analyzed natural monopoly aspects of DH [30-32]; pricing of DH in different countries [33-35]; regulatory issues of DH [36, 37]. Authors [38] analyzed economic regulated pricing that is prevalent in countries after reforms without well-established infrastructure, where dominates strong monopolies in energy sectors.

The main reason why DH in transition economies has lower efficiency than in Western Europe is not only regulation [39]. Favorable energy policy and support for the use of RES are insufficient due to high prices of alternative energy technologies. Therefore, total RES development is not impressive.

GUIDELINES FOR PROMOTION OF RENEWABLE ENERGY TECHNOLOGIES

Integration of RES Technologies into Energy Development Process: A Local Approach

The most important question of our research is the evaluation of progress of reliability and other sustainable development areas on the territorial aspect. The territorial dimension means implementation of management principle while evaluating sustainability components of energy resources in line with defining the limits of liability.

"Throughout the world, increasing attention is being paid to climate change mitigation. But while ambitious national targets are hard to come by, several regions, cities, towns, institutions, and individuals have taken matters into their own hands. Rather than awaiting international agreement or national targets, these established their own ambitious targets for reducing carbon dioxide emissions and are in the process of finding ways and means to meet these targets. Various levels serve as impetuous for more ambitious targets by demonstrating a will to policymakers as well as finding and demonstrating feasible options. It helps confront the previous paradigm that fossil-fueled economies are the only viable options and renewable energy systems cannot supply the required energy services at a competitive cost" [40].

According to Klessmann et al. [24], "Critical success factors include implementing effective and efficient policies that attract sufficient investments, reducing administrative and grid related barriers, especially in currently less advanced countries, dismantling financial barriers in the heat sector, realizing sustainability standards for biomass, and lowering energy demand through increased energy efficiency efforts".

A very important question is a selection of research models and feasibility analysis. EnergyPRO is typically used for techno-economic analysis of simulating cogeneration plants and DH systems with multiple energy producers [21, 22, 41]. Other types of projects, e.g., solar collectors and heat pumps, can also be analyzed and detailed within the software [42].

The increasing complexity of management of the quickly developing cities in Europe requires an integrated approach, which facilitate planning and developing departments of the city council to continue sustainable development on a city level [43-45].

"When cities present proper densities and planning, they demonstrate to be the most efficient way of life. Then the recognition of the necessary accountability of cities is clear. Because most environmental problems that society faces today have their origin in urban areas, they must combine commitment and capacity for innovation to solve them" [46]. The local government of each city together with society should find an individual way of sustainable development because cities are different in their size of the territory, a number of inhabitants, environment, political and social-cultural conditions. The most important thing is the participation of city inhabitants, representatives of business and other sectors in urbanized life aspects because cities, in a sense, are products of their inhabitants [47].

The decisions mainly aim at the short-term competitiveness of price in the heating sector, without evaluation of long-term benefits or possibility to stabilize the price of heat, therefore, the cheapest source of fuels is dominant. This is explained by the fact that the uptake of other types of RES (solar, geothermal) requires in-depth research.

"There is a growing interest in the geothermal resources available at shallow depth beneath cities. However, there exists no general procedure for quantifying the low-temperature geothermal potential in urban ground and groundwater. The broad span of the results highlights the need for a more consistent framework that distinguishes between the conceptual assumptions for calculating the technical geothermal potential and the local city-specific factors. This will be the basis for a reliable analysis of the economic geothermal potential of low-temperature geothermal applications on a local, district or city scale. This will also enhance the reliability and the trust in these technologies, and thus the public acceptance reflected in the acceptable geothermal potential" [48].

Many examples of the worldwide practice have shown how cities and villages on the territorial basis consolidate and manage funds, including private finances, which are allocated or invested in renewable energy technologies. Particularly noteworthy is the cooperation between urban communities and the private sector in financing RES projects, which was analyzed in [49]. Authors' [49] review describes the fast-growing and significant phenomenon of community and private sector partnerships in renewable energy by identifying six archetypes: knowledge sharing, private finance, local consumption, land seeking, community employment, and lease. These archetypes are discussed with relation to key variables influencing partnership formation and perpetuation as part of the wider discussion regarding renewable energy partnerships between local communities and the private sector and the role they play in the energy transition to a low-carbon society.

Substantial obstacles (except financing issues) to take appropriate actions in municipalities are not existent. Some of them have signed the Covenant of Mayors but only a few take active steps.

In addition to broad responsibilities for the preparation of planning documents, during the past few years the role of local self-governments significantly enhanced in apartment building renovation (modernization).

Figure 2 shows principles for performing tasks of national importance, which are related to sustainable energy development (the top part of the scheme), and how it can be solved on the territorial principle. Connecting link for the practical implementation of national RES energy goals should be municipalities. The role of municipalities in the development of RES is regulated in detail in the Law on Energy from Renewable Sources. However, the implementation of this law on the municipal level does not progress in practice. RES Development Action Plans have been developed only by few municipalities, but this claim at municipal councils has stalled in Lithuania. According to the existing situation, it can be stated that problems and failures in case of municipal involvement in the planning of RES refer to the short-term and fragmented solutions.

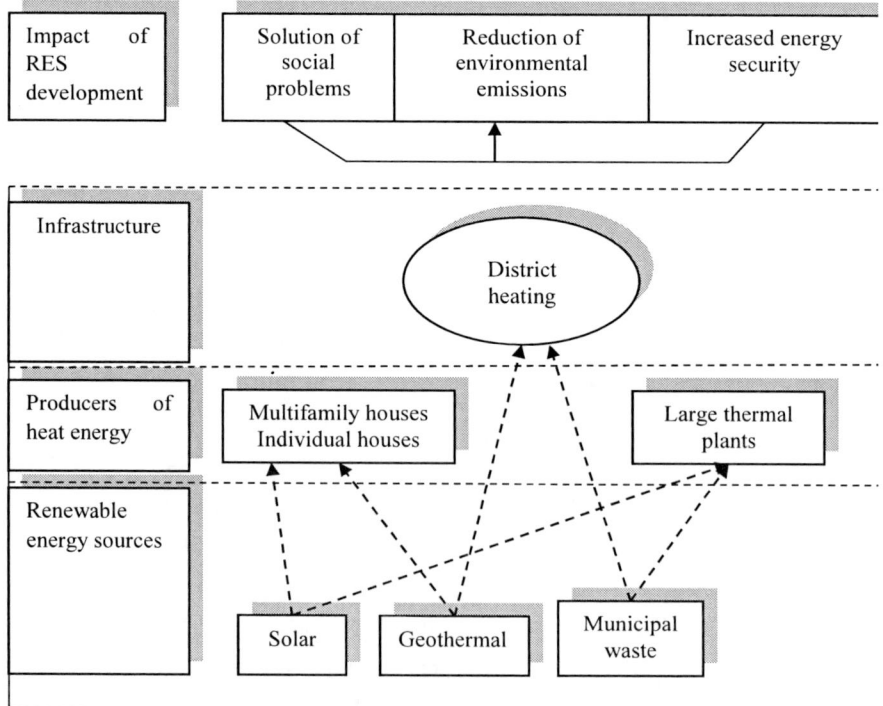

Figure 2. Significance of district heating infrastructure for solving national sustainable development problems on territorial principles.

In the previous section, it was explained that various programs, which are mandatory for the jurisdiction of cities according to new laws, should be merged into a uniform program, whilst providing stages of its implementation and sufficient funding after a consolidation of various funding sources. Figure 2 shows that a special role has the use of DH infrastructure.

It should also be noted that the district cooling supply also has its advantage because its infrastructure, as in the case of district heating, allows integrating different types of RES.

"District cooling system (further in the text DCS) is a centralized system of supplying the thermal energy in form of chilled water for the use in space cooling and dehumidification. In addition, DCS integrated with renewable energy is economically feasible when compared to conventional cooling systems. The most suitable renewable energy technologies that can be integrated with DCS are biomass energy, solar thermal energy, geothermal energy, surface water energy, solar photovoltaic energy, and waste heat energy" [50].

Strategic energy goals could be implemented by using DH and DCS infrastructure that is favorable for integrating RES, using municipal waste for combined heat and power (further in the text CHP), increasing the energy security by diversified fuel sources, implementing DCS for public and private buildings, using of industrial waste for heating, simultaneous producing of heat and electricity in cogeneration process, reducing air pollution in cities and mitigating climate change, etc.

DH infrastructure should be used for diversification of potential fuel sources, thus increasing security of energy supply and long-term stabilization of energy prices.

Lithuanian DH sector is focused on the use of biomass that has a dominant position due to the lowest price in the short-term period. Therefore, diversification of RES, such as solar and geothermal energy, is limited in very well-extended infrastructure of DH.

One of the reasons from an economic perspective is that DH systems are not treated as natural monopolies, and pricing, which encourages production and supply of the maximum amount of heat energy. This is a consequence of wrong treatment and the main source of economic problems in DH systems.

Another issue is a mismatch between the interests of consumers and producers in DH. During the evaluation process of the use of RES in multifamily buildings, support for RES and possible consequences for the DH infrastructure must be evaluated. There is no doubt that residents receiving subsidies for renovation of multifamily buildings and for development of RES will affect large savings of heat energy. Therefore, it may cause a significant decrease in heat consumption and encourage massive disconnections from the DH system, which will not only imbalance the DH network but also raise the price for the rest of consumers. The loss in the DH network should be paid by the lower number of heat consumers. The massive renovation of multifamily buildings will reduce the amount of consumed heat energy significantly; for this reason, the price for DH will increase noticeably.

Research and decisions should be based on an economic approach for DH as the natural monopoly. DH infrastructure or heat distribution system is a natural monopoly, thus it has specific pricing, regulation, and infrastructure planning. Examples of north European countries in similar climate zone show that the problems could be successfully solved. It should be noted that limitations of monopoly can be addressed by any means which lead to natural monopoly status. One of the most important preconditions is territorial planning of energy development; implementation of plans could implement fuel diversification in the heating sector on the planning level in order to avoid excess investment.

Environmental issues have reached alarming proportions. Therefore, for the first time in history, the Paris Agreement has been adopted unanimously by all parties. The implementation of the Paris Agreement requires the substitution of combusted fossil fuel by RES. It confirms the latest economic theory about the circular economy, which vital importance for humanity is undoubted, especially when business attitudes towards nature is like a warehouse of raw materials. The concept of a circular economy process is based on local and renewable energy resources as well as the application of zero waste technologies. The main objectives are to use local resources of every region, create a unique environment for the use of these resources, and form local industry and agriculture by estimating the ecological and social value of such economic development. The aim of conservation of nature and heritage of culture and history is in the concept of sustainable development, as well as in most of the XXI century ideas about the future of the new way of life. The example of practical implementation of such ideas is eco-villages movement. The eco-village is local initiatives and activities of group leaders. The main purpose is living in a clean and ecological environment, fostering a sense of community and social unity, and promoting the development of organic farm. According to the international organization Global Ecovillages Network, nowadays more than 40 thousand eco-villages exist in more than 40 countries.

Experience has shown that eco-villages movement solves problems of employment, business promotion, social exclusion, and alternative business creation in rural and suburban areas. Special attention in these communities is given to the use of RES.

In the world and especially in Europe there are long-term programs on a large scale in order to use exclusively RES for energy supply in cities and regions.

Table 2 shows a list of cities that are committed over a period of time reaching 100 percent supply of energy from RES. There were a number of attempts to move in this direction. It might seem that it requires billions of funds. However, the circular economy theory and facts that are supporting this theory shows that enhanced support and funds are necessary for the initial stage, especially during the formation process of awareness raising systems. On the other hand, respect for the environment is rarely considered as the unconditional value of business relations. Educational activities are necessary, which would show the connection between human activity and eternal laws of nature. A special role in this process is attributed to smaller cities and rural or suburban communities. It will be a long process of transformation, but this is one of two alternatives: the continued destruction of the natural environment or the circular economy that is based on the above principles.

Table 2. Countries and cities which are planning 100% RES in different continents [51]

Europe	Number of cities, regions	Knowledge dissemination	Solar PV	Solar Collectors	Wind	Biogas	Geothermal	Smart grids	Biomass	Hydro	Electric vehicles
Austria	7	+	+		+	+		+	+	+	
Denmark	9	+	+		+	+	+	+			+
Germany	40	+	+	+	+			+	+	+	+
Belgium	1	+						+		+	
France	2	+	+	+	+			+			
Turkey	1	+	+		+					+	
Italy	5	+	+	+	+	+		+	+	+	
Switzerland	2	+	+					+	+		
Netherlands	2	+			+			+			
Iceland	2	+					+	+		+	
Poland	1	+	+		+			+	+		
Portugal	1	+	+		+			+	+	+	
Slovenia	1	+							+		
Spain	2	+	+	+	+					+	
Sweden	4	+	+		+	+		+	+		+
United Kingdom	4	+	+		+			+	+	+	

These initiatives pale in comparison with the ambitions of Denmark as a state to achieve 100 percent self-sufficiency of RES in all sectors by 2050. However, this requires very intensive educational and organizational activities of local authorities for synchronizing the implementation of state objectives and funding and incentive mechanisms with real opportunities of cities and appropriate regulatory scheme.

Transformation of the DH System towards More Diversified Renewable Technologies in the Fuel Balance of Heat Sector

Nowadays the main goal for the use of DH energy is the cheap price of heat for the final consumers, while for the supplier of DH it is financial self-sufficiency in obtaining profits. DH companies now operate as profit-seeking enterprises, but due to monopoly position, it has exclusive rights, such as an adequate rate of return and calculation of price on a monthly basis.

Firstly, it should be defined what is the progress of the DH sector. Apriori it is considered that the DH systems in Lithuania are well developed. Physically, according to potential opportunities of DH, it is an advanced heating system. However, the comparison with foreign countries shows that such DH systems require further development.

The analysis of Denmark and other countries, especially in Northern Europe, highlighted several key points. Firstly, it is a stable heat price for consumers in the long-term, which is achieved by the fuel that contains a considerable diversification of RES. Secondly, it is constantly developing DH system, which is already the fourth generation of technological structure.

This section examines direction towards more diversified renewable technologies in the fuel balance of DH system. It should be noted that reforms of the DH sector cover a wide range of related technical, legal, economic and organizational measures, which has its own aspects in individual countries. However, the basis of all reforms and improvement of technologies is the fact that DH is a natural monopoly, rather than the market economy enterprise.

The Scandinavian countries, as well as Lithuania, have a well-developed DH infrastructure and long-term experience of planning and legal regulation.

Lithuanian DH sector has been facing challenges not only during recent years. After the adoption of the Law on Energy from Renewable Sources [52] and EU Structural Support period 2007-2013, there was no stable state policy on DH. Support was given for small scale biomass boilers up to 5 MW capacity and competition on the production side of DH was encouraged by the Law on Heat Sector [53]. On the other side, pricing was not favorable for municipalities' producers, especially in large Lithuanian cities; therefore, many small private investors started the business at producing heat in small biomass boilers. As an example of successful energy policy and strategic planning could be Denmark.

According to Sperling et al. [54], the transition period of energy system based on fossil fuel technologies and centralization in Denmark started since the 1970s. One of the main results was long-term energy planning, which had been influenced by an open and flexible political process. The introduction and dissemination of cogeneration (CHP) technology as well as DH, wind energy, and biogas were priorities for support from targeted programs. Denmark's first overall energy plan "Danish energy plan" was enacted in 1976. The plan was intended to lay the basis for a long-term energy policy. The public policy in Denmark has a regulatory approach with a specific heating law, including tariff regulation, zoning, etc. DH has a strong position in the Danish heat market. As demonstrate Danish Energy Agency [55], two major issues in the Danish law (Act on Heat Supply) have large effects in the market. First, since 1982 there is an obligation to connect new and existing buildings to public supply. Second, a ban on installing electrical heating in new buildings is since 1988 and a ban on installing electrical heat systems in existing buildings with water-based central heating since 1994. In comparison to Lithuania, there is no existing ban for electric heating or the use of natural gas in buildings; therefore DH should compete with alternatives on the market conditions.

An important role is given to municipalities in Denmark. As stated by Nielsen and Meyer [56], heat planning was introduced in 1979, and it was mandatory for municipalities to establish a heat plan in accordance with specific rules given by the Ministry of Energy. Strategic municipal heat planning was successful in the 1980s but became less effective after the revision of the Act on Heat Supply in 1990, as argue Sperling and Möller [57]. The first Danish Act on Heat Supply of 1979 had the requirement to designate geographical heat supply areas; responsibility was given to the municipalities together with counties, and local utility companies. As a result, heat supply made a transition from individual forms of heating, such as oil boilers or electric heating, to more efficient and collective forms, such as DH and natural gas. After the establishment of well-defined heat supply areas, most municipalities restricted themselves to acting as project authorities that would approve single heat supply projects without having long-term aspects in mind. Municipalities still have the authority to devise heat plans, which nowadays mainly contain general objectives and can be the basis for initiating specific heat supply projects. In general, overall municipal heat planning has been given a lower priority in many municipalities during the last two decades. In Lithuania, municipality role in the DH sector is usually defined by the decision of DH price in small municipalities for a 3-5 years period. Zoning of DH areas in cities/towns of Lithuania is still rare. Decisions, however, also requires clear support and guidelines from the central level, which should set the frame for strategic municipal energy planning in general, and municipal heat supply and demand planning in particular.

Nielsen and Möller [58] state that since the late 1970s Danish energy administration has been using a geographical database for heating demand and supply. Authors express the opinion that the geographical database could be a base for the future stabilization of state policy for DH. After each change of government, there is a tradition in Lithuania to change National Energy Strategy, there is no available geographical database. A lack of stable strategic planning and a more active role of municipalities are the key elements that should be improved in the nearest future.

Denmark's price regulation on DH was introduced with the Act on Heat Supply of 1979. The fact that DH in Denmark supplies a large number of individual houses, together with the option of mandatory connection, gives it a dominating position in the heat market and natural monopoly. Therefore, the act specifies that DH must be operated as a non-profit activity, with cost-based pricing. Nowadays in Denmark DH is owned by consumer cooperatives or municipal utilities. On the contrary, in Lithuania DH companies are mainly owned by municipalities or private investors.

Sweden has an extensive DH sector. DH accounts for about 40% of the heating market in Sweden. The change in the fuel mix has been impressive: compared to 1970, when oil was the main fuel, oil accounts for only a few percent today. More than 62% of DH fuel today is biomass [59]. The main key forces for the transition from oil to biomass are favorable public policies, such as the introduction of a carbon tax in 1991 and the Tradable Renewable Electricity Certificates scheme in 2003. Compared to Lithuania, the main force to make the transition from natural gas based DH to biomass-based DH is support from EU funds and requirements of RES Directive. The lack of long-term energy policy creates the situation when support is concentrated only till 2020 in Lithuania.

The Swedish DH market is already deregulated in the sense that companies may charge the price they see fit (unless they are municipal administrations). It is also deregulated in the meaning that any company wishing to enter the market by starting a DH business may do so, provided relevant permits, etc. granted by authorities. On the other hand, DH is not deregulated since all companies are vertically integrated and when a local de-facto monopoly for DH has been established no competition exists in the specific DH market [60]. In Lithuania, DH production should work on market conditions but is still regulated by the National Commission for Energy Control and Prices (NCECP).

Denmark and Sweden are small and have a well-developed DH infrastructure. Lithuania was part of the Soviet Union, and DH was a monopoly during those times. Nowadays a competitive market in the DH sector is encouraged by Law on Heat Sector and support is given mainly from EU Structural Funds.

Environmental benefits of DH are integrated as a part of public policy in Denmark and Sweden. On the other hand, Finland has chosen less sustainable development of DH.

DH has a stable pricing policy in Finland. Most fuel prices have risen steeper than the price of DH since 2004. One reason why the DH sector has managed this well is the diversified use of fuels, another is the increase in the use of peat and waste wood, and a third is the widespread use of coal [61]. The prices of fossil fuels increased after 2011 due to implemented environmental taxes in Finland. Finland, as well as Denmark and Sweden, use high environmental taxes for fossil fuel with the exception for CHP. Lithuania is not going the same direction at the moment; on the contrary, the feed-in tariff for electricity from CHP has decreased during recent years.

As reported by Wessberg [62], DH production originated in Finland in the 1950s. Following the initiative, integrated heat and power production has become a significant element of energy production in Finland. Centralized heat production using water boilers started simultaneously with integrated production. DH almost became a fashion in Finnish municipalities during the 1970s after the oil crisis. The most encouraging results from the viewpoint of policy are the deregulation of the market and restructuring of the industry. The restructuring of the electricity market since 1999 had an indirect impact on lower DH prices [16]. In Lithuania, the lower feed-in tariff for electricity from CHP also had an indirect effect on DH prices, but negative.

To summarize, Denmark has heat zoning, many decentralized DH plants, and working on a non-profit basis. RES integration is based on taxation of fossil fuels; therefore, the preeminent position of RES is even without the support for investment or operation. Sweden has chosen a deregulated market of DH but has influenced the market by taxation for fossil fuel. Finland has chosen a less sustainable DH sector, which is based on diversification (both RES and fossil fuel). The preeminent position is given to the lower price. All three countries have high energy and environmental taxes.

Denmark was a pioneer country which chosen priority for energy savings and diversification of energy supply, including the use of RES. Innovative energy-policy initiatives were focused on CHP, heat planning on the municipal level and development of the extensive natural gas grid. Moreover, the energy efficiency of building stock was implemented, support for RES was given, research of environmentally friendly energy technologies was enhanced, and ambitious green taxes were introduced. Denmark made a transition from oil importing country in the 1970's to more energy self-sufficient country at the end of the XX century. The focus of energy policy on the increase of RES together with a continuous reorganization of energy supply was a basis for Denmark enabling ambitious goals for the reduction of greenhouse gas emissions and 100 percent of RES in 2050. Comparison of Denmark and Lithuania long-term policy of DH is presented in Table 3.

Table 3. Comparison of Denmark and Lithuania long-term policy of district heating

	Denmark	Lithuania
Long-term strategy	100 percent of RES in 2050	23 percent of RES for total consumption of energy in 2020
Tax system RES support system - For producers - For consumers	Taxes for fuels Programs of the cities	No taxes for fuels Support measures are aimed at implementation of ES directives
Infrastructure	District heating	District heating
Interaction between consumers and producers	Scenarios methods Non-profit organizations Regulated by antimonopoly law Binomial tariff	Profit seeking organizations Monomial tariff, controlled by NCECP

According to OECD [63], gas prices in Denmark before taxes are similar to other EU countries; however, the final retail price of gas in Denmark is very high among OECD members because of high taxes. Households had a burden of 50.6% of taxes of natural gas prices in 2010. Consumers of DH pay a lower fee for energy from CHP plants in Denmark.

In conclusion, the natural preeminent position of RES in Denmark DH system is created by long-term energy policy, which incorporates externalities of fossil fuel into price under different taxes. Therefore, non-exhaustible RES could compete on market conditions even without intervention from the state. State policy plays a key role in decision-making of the country for the future perspectives on the use of diversified RES sources.

Long-term political ambitions in renewable energy technology utilization are shown in Table 4. Lithuania is in the second generation of DH. The use of heat storage and large scale solar collectors are just a theoretical possibility at the moment.

Table 4. Evolution of district heating in Denmark and Lithuania (3 generations) [26]

Basic system parameters	Steam systems, steam pipes in concrete ducts	Pressurized hot water system, heavy equipment, substations		Pre-insulated pipes, compact substations, metering and monitoring, low energy buildings	
Country	Denmark	Denmark	Lithuania	Denmark	Lithuania
Temperature	<200^{0}C	>100^{0}C	>100^{0}C	70<100^{0}C	<100^{0}C
Structure of generating technologies	Local district heating Steam storage Coal, waste Heat storage CHP coal CHP oil	District heating Large scale solar Biomass CHP biomass Industry surplus Heat storage CHP waste incineration CHP coal CHP oil Gas boilers Waste boilers Oil boilers Coal boilers	District heating Oil boilers Gas boilers CHP oil CHP gas	District heating Seasonal heat storage Large scale solar Geothermal Wave Wind surplus Electricity Heat storage Industry surplus CHP waste incineration Low energy buildings	District heating Biomass boilers Gas boilers CHP gas CHP waste incineration Geothermal Low energy buildings
DH generation	1G/1880-1930	2G/1930-1980	1G/1947-1990	3G/1980-2020	2G/1990-2020

The use of RES in production of DH energy would allow diversifying the fuel and energy sources. Moreover, the use of diversified RES in the production of heat energy must be enhanced by other measures than existing. Feed-in tariff example for solar PV expansion in Lithuania might be an example of the power of support measures for immature technologies to take part in the market. Moreover, declining of solar PV, called a Swanson's law, predicted 20% price declines for every doubling of installed capacity. The similar effect is expected to solar collectors due to the high impact of China, which produces a major part of solar collectors.

A management aspect is not less important. The territorial authorities of state government, such as municipalities or their energy development agencies, have to be managers and organizers that are responsible for the implementation of territorial district, city, region, and at the same time state obligations.

The comparison with the case of Denmark might show possible directions for the further development of the DH sector by the use of diversified RES.

CONCLUSION

Renewable energy policy development is a complex system, where a balance of three aspects of sustainability is needed: environment, economy, and social life. Our generation has a challenge not only to cope with problems of energy supply without leaving them to other generations to solve it, but especially with the growing threat of climate change. Deployment of renewable energy technologies receives special attention because the burning of fossil fuel is identified as the main cause for pollution. However, good performance and well-balanced renewable energy policy require efforts from stakeholders with different interests and market participants.

It is recognized that integrated complex of environmental, social, and economic policies is easier implemented in cities and districts. Public expenditure and funding from Structural Funds should, where possible, allow achieving more than one policy objective of sustainable development. Sustainable development of energy on the regional level requires institutions to organize and implement support measures. Municipalities could play an important role in promoting assumptions of sustainable energy development because the local government performs a number of functions in the energy sector, and these opportunities should be used.

Legal obstacles for municipalities to plan renewable energy development and energy efficiency measures are non-existent in laws.

Infrastructure of district heating and district cooling systems should be used for implementing general objectives of the country, such as security of energy supply, reduction of greenhouse gas emissions, diversification of fuel mix in order to use local and renewable energy resources, efficient generation of electricity from cogeneration, utilization of municipal waste for generation of heat and electricity, integration of industrial waste into district heating networks.

The analysis of the Lithuanian energy system showed that responsibilities of local self-governments for renewable energy development are sufficiently regulated. However, the main problem is that implementation of individual EU directives for local authorities is obliged by creating separate programs which do not have a unified consolidation. Most local self-governments delayed the implementation of statutory renewable energy programs and plans. Scarce resources at the disposal of municipalities are scattered. Our recommended scheme for programs and funds consolidation would accelerate the process of sustainable energy.

The massive use of RES is a global challenge, but its solution is implemented locally. That is impeded not only by a variety of technological obstacles but also by unorganized individual and often contradictory knowledge, which are disseminated for advertising purposes, concealing the negative consequences for the public and the damage to nature. Because of the individual business approach it is natural to take full advantage of the offered opportunities; however, public interests are often infringed, i.e., people who are not involved in a business transaction. Because the goal of municipalities is the welfare of society, municipal authorities can concentrate the whole spectrum of knowledge for the implementation of a large scale of RES.

Developed Western countries are a good example how ecological balance can be maintained: biodiversity combining various RES technologies, existing infrastructure of district heating supply, development of smart technologies that can transform consumers into energy producers (prosumers).

REFERENCES

[1] Ciegis, R. 2002. *Continuous development and environment: an economic approach* [in Lithuanian]. Vilnius: ISM University of Management and Economics.

[2] Cruz, M.R., Fitiwi, D.Z., Santos, S.F, and Catalão, J.P. 2018. "A comprehensive survey of flexibility options for supporting the low-carbon energy future." *Renewable and Sustainable Energy Reviews* 97:338–53. doi:10.1016/j. rser.2018.08.028.

[3] Shen, Y.C., Lin, G.T., Li, K.P., and Yuan, B.J. 2010. "An assessment of exploiting renewable energy sources with concerns of policy and technology". *Energy Policy* 38:4604–16. doi:10.1016/j. enpol. 2010.04.016.

[4] Marques, A.C., Fuinhas, J.A., and Pires Manso, J.R. 2010. "Motivations driving renewable energy in European countries: A panel data approach." *Energy Policy*, 38:6877–85. doi:10.1016/j. enpol.2010.07.003.

[5] Katinas, V., Markevicius, A. 2006. "Promotional policy and perspectives of usage renewable energy in Lithuania." *Energy Policy* 34:771–80. doi:10.1016/j. enpol.2004.07.011.

[6] Koch, H.J. 2012. *Accelerating renewable energy technology deployment.* Renewable Energy Action on Deployment.

[7] European Council. Directive 2009/28/EC of the European Parliament and of the Council on the *promotion of the use of energy from renewable sources and amending and subsequently repealing Directives 2001/77/EC and 2003/30/EC.* Off. J. Eur. Communities, 5.6.2009, L 140/16, 2009.

[8] Hudson, R. 2011. "From knowledge-based economy to ... knowledge-based economy? Reflections on changes in the economy and development policies in the North East of England." *Reg Stud* 45:997–1012. doi:10.1080/00343400802662633.

[9] Guide for Awareness-Raising Campaigns. 2012. *Global Solar Water Heating Market Transformation and Strengthening Initiative.* The European Solar Thermal Industry Federation (ESTIF), the United

Nations Environment Program (UNEP) through its Division of Technology.

[10] *Education on Energy. Teaching tommorow's energy consumers. 2006.* European Commission and Directorate General on Energy and Transport, Luxembourg: Office for Official Publications of the European Communities.

[11] Tamale, E. 2011. *Introduction to public awareness and education: concepts, communication strategies and messaging techniques. Convention on Biological Diversity 2011-2020.* United Nations.

[12] *Building a Sustainable Energy Future.* 2009. National Science Board Foundation.

[13] Jennings, P. 2009. "New directions in renewable energy education." *Renew Energy* 34:435–9. doi:10.1016/j. renene.2008.05.005.

[14] Camagni, R. 2002. On the concept of territorial competitiveness: sound or misleading? *Urban Stud* 39:2395–411.

[15] Tanguay, G., Rajaonson, J., Lefebvre, J., and Lanoie, P. 2010. Measuring the sustainability of cities: An analysis of the use of local indicators. *Ecol Indic* 10:407–18.

[16] Linden, M., and Peltola-Ojala, P. 2010. "The deregulation effects of Finnish electricity markets on district heating prices." *Energy Econ* 32:1191–8. doi:10.1016/j. eneco.2010.03.002.

[17] Ciegis, R., and Streimikiene, D. 2005. "Integration of sustainable development indicators in development programmes." *Econ Eng Decis* 42:7–13.

[18] Ciegis, R., Grundey, D., and Streimikiene, D. 2005. "Economic aspects of cities sustainable development strategic planning." *Technol Econ Dev Econ* 11:260–9.

[19] Streimikiene, D., and Pareigis, R. 2007. "Promotion of use of renewable energy sources in Lithuania." *Technol Econ Dev Econ* 13:159–69. doi:10.1080/13928619.2007.9637792.

[20] Lund, H., Siupsinskas, G., and Martinaitis, V. 2005. "Implementation strategy for small CHP-plants in a competitive market: the case of Lithuania." *Appl Energy* 82:214–27. doi:10.1016/j. apenergy. 2004.10.013.

[21] Rasburskis, N., and Lund, H. 2007. "Optimization methodologies for national small-scale CHP strategies (the case of Lithuania)." *Energetika* 53:16–23.

[22] Streckiene, G., Martinaitis, V., Andersen, A.N., and Katz, J. 2009. "Feasibility of CHP-plants with thermal stores in the German spot market." *Appl Energy* 86:2308–16. doi:10.1016/j. apenergy. 2009.03.023.

[23] Konstantinaviciute, I. 2011. "Policy assessment: Case study Lithuania. *Paper presented at RES-H Policy Proj. Natl. Dissem. Conf.* Vilnius, Lithuania.

[24] Klessmann, C., Held, A., Rathmann, M., and Ragwitz, M. 2011. "Status and perspectives of renewable energy policy and deployment in the European Union—What is needed to reach the 2020 targets?" *Energy Policy* 39:7637–57. doi:10.1016/j. enpol.2011.08.038.

[25] Marcinauskas K, and Korsakienė I. 2011. "District heating and heat prices 1945–2011 in Lithuania: historical-expert review." *Energy* 57:207–30. [in Lithuanian]

[26] Lund, H., Werner, S., Wiltshire, R., Svendsen, S., Thorsen, J.E., and Hvelplund, F. 2014. "4th Generation District Heating (4GDH). Integrating smart thermal grids into future sustainable energy systems." *Energy* 68:1–11. doi:10.1016/j. energy.2014.02.089.

[27] Gadd, H., and Werner, S. 2014. "Achieving low return temperatures from district heating substations." *Appl Energy* 136:59–67. doi:10.1016/j. apenergy.2014.09.022.

[28] Mathiesen, B.V., Lund, H., Connolly, D., Wenzel, H., Østergaard, P.A., and Möller, B. 2015. "Smart energy systems for coherent 100% renewable energy and transport solutions." *Appl Energy* 145:139–54. doi:10.1016/j. apenergy.2015.01.075.

[29] Brand, L., Calvén, A., Englund, J., Landersjö, H., and Lauenburg, P. 2014. "Smart district heating networks – A simulation study of prosumers' impact on technical parameters in distribution networks." *Appl Energy* 129:39–48. doi:10.1016/j. apenergy.2014.04.079.

[30] Depoorter, B.W. 1999. "Regulation of natural monopoly. In: Bouckaert, B., De Geest G, editor." *Encycl. Law Econ, Edward Elgar*, Cheltenham; p. 498–532.

[31] Mosca, M. 2008. "On the origins of the concept of natural monopoly: Economies of scale and competition." *Eur Hist Econ Thought* 15:317–53. doi:10.1080/09672560802037623.

[32] Magnusson, D., and Palm, J. 2011. *Between natural monopoly and third party access - Swedish district heating market in transition.* New York: Nova Science Publishers.

[33] Li, H., Sun, Q., Zhang, Q., and Wallin, F. 2015. "A review of the pricing mechanisms for district heating systems." *Renew Sustain Energy Rev* 42:56–65. doi:10.1016/j. rser.2014.10.003.

[34] Difs, K., and Trygg, L. 2009. "Pricing district heating by marginal cost." *Energy Policy* 37:606–16. doi:10.1016/j. enpol.2008.10.003.

[35] Björkqvist, O., Idefeldt, J., and Larsson, A. 2010. "Risk assessment of new pricing strategies in the district heating market." *Energy Policy* 38:2171–8. doi:10.1016/j. enpol.2009.11.064.

[36] Wissner, M. 2014. "Regulation of district-heating systems." *Util Policy* 31:63–73. doi:10.1016/j. jup.2014.09.001.

[37] Lukosevicius, V., and Werring, L. 2011. *Regulatory Implications of District Heating.* Energy Regulators Regional Association (ERRA).

[38] Aronsson, B., and Hellmer, S. 2009. An International Comparison of District Heating Markets. *Rapport Svensk Fjarrvarme.* ISBN 978-91-7381-042-5.

[39] Poputoaia, D., and Bouzarovski, S. 2010. "Regulating district heating in Romania: Legislative challenges and energy efficiency barriers." *Energy Policy* 38:3820–9. doi:10.1016/j. enpol.2010.03.002.

[40] Ostergaard, P.A., and Lund, H. 2010. "Climate Change Mitigation from a Bottom-up Community Approach." *Sustainable communities design handbook* https://doi.org/10.1016/B978-1-85617-804-4.00014-8.

[41] Fragaki, A., and Anderse, A.N. 2011. "Conditions for aggregation of CHP plants in the UK electricity market and exploration of plant size." *Appl Energy* 88:3930–40. doi:10.1016/j. apenergy.2011.04.004.

[42] Nielsen, S., and Möller, B. 2012. "Excess heat production of future net zero energy buildings within district heating areas in Denmark." *Energy* 48:23–31. doi:10.1016/j. energy.2012.04.012.

[43] Xing, Y., Horner, R., El-Haram, M., and Bebbington, J. 2009. A framework model for assessing sustainability impacts of urban development. *Account Forum* 33:209–24.

[44] Rotmans, J., and Asselt, M. Van. 2000. Towards an integrated approach for sustainable city planning. *Multi-Criteria Decis Anal* 9:110–24.

[45] Walton, J., and El-Haram, M. 2005. Integrated assessment of urban sustainability. *Eng Sustain* 158:57–65.

[46] Pereira, J., and Azevedo, A. 2011. Interdependency between sustainable development and economic growth (investment attraction): the role of city's governance, branding and monitoring. *J Mod Account Audit* 7:734–48.

[47] Ciegis, R., and Cesonis, G. 2004. "Sustainable development strategic planning: urban aspect." *Strateg Self-Management*, 20–31.

[48] Bayer, P., Attard, G., Blum, P., and Menberg, K. 2019. "The geothermal potential of cities." *Renew Sustain Energy Rev* 106: 271-280. doi.org/10.1016/j.rser.2019.02.019).

[49] Eitan, A., Herman, L., Fischhendler I., and Rosenc, G. 2019. "Community–private sector partnerships in renewable energy." *Renew Sustain Energy Rev* 105: 95-104. doi.org/10.1016/j.rser. 2018.12.058.

[50] Inayat, A., and Razab, M. 2019. "District cooling system via renewable energy sources: A review." *Renew Sustain Energy Rev* 107: 360-73. doi.org/10.1016/j.rser.2019.03.023.

[51] Go 100% *Renewable Energy 2016*. http://www.go100percent.org/cms/index.php?id=19(accessed December 16, 2016.

[52] Lithuanian Republic. 2011. Law on Energy from Renewable Sources of the Republic of Lithuania. *Official Gazette* No. 62-2936.

[53] Lithuanian Republic. 2003. Law on Heat Sector of the Republic of Lithuania. *Official Gazette* No. 51-2254.

[54] Sperling, K., Hvelplund, F., and Mathiesen, B.V. 2011. "Centralisation and decentralisation in strategic municipal energy planning in Denmark." *Energy Policy* 39:1338–51. doi:10.1016/j.enpol.2010.12.006.

[55] Danish Energy Authority. 2005. *Heat supply in Denmark*. Copenhagen: Danish Energy Authority.

[56] Niels, I., and Meyer, B.V. 2014. Barriers and Potential Solutions for Energy Renovation of Buildings in Denmark. *Sustain Energy Plan Manag* 01:59–66.

[57] Sperling, K., and Möller, B. 2012. "End-use energy savings and district heating expansion in a local renewable energy system - A short-term perspective." *Appl Energy* 92:831–42. doi:10.1016/j.apenergy.2011.08.040.

[58] Nielsen, S., Möller, B. 2013. "GIS based analysis of future district heating potential in Denmark." *Energy* 57:458–68. doi:10.1016/j.energy.2013.05.041.

[59] Di, L., and Ericsson, K. 2014. "Low-carbon district heating in Sweden – Examining a successful energy transition." *Energy Res Soc Sci* 4:10–20. doi:10.1016/j.erss.2014.08.005.

[60] Westin, P., Lagergren, F. 2002. Re-regulating district heating in Sweden. *Energy Policy* 30:583–96.

[61] Havskjold, M., and Sköldberg, H. 2009. *The future of the Nordic district heating. A first look at district heat pricing and regulation.* Oslo: The NEP research group.

[62] Wessberg, N. 2002. "Local decisions in the Finnish energy production network - A socio-technical perspective." *Landsc Urban Plan* 61:137–46. doi:10.1016/S0169-2046(02)00108-1.

[63] OECD 2013. *Inventory of Estimated Budgetary Support and Tax Expenditures for Fossil Fuels.* Paris: OECD.

In: A Comprehensive Guide to Energy ...　ISBN: 978-1-53616-728-3
Editor: Shannon Alvarado　© 2020 Nova Science Publishers, Inc.

Chapter 2

TRANSFORMATION PATHWAYS TOWARDS A CLEAN, SECURE AND EFFICIENT EUROPEAN ENERGY SYSTEM: A MCDA APPROACH

Charikleia Karakosta[*], *PhD and Aikaterini Papapostolou*
Decision Support Systems Laboratory (EPU-NTUA)
School of Electrical and Computer Engineering,
National Technical University of Athens,
Zografou, Athens, Greece

ABSTRACT

The European Union's (EU) energy, innovation and climate challenges define the direction of a future European energy system, but the specific technology pathways are policy sensitive and require careful comparative evaluation. The Strategic Energy Technology Plan (SET-Plan), as part of the technology pillar of European Energy and Climate

---

[*] Corresponding Author's E-mail: chkara@epu.ntua.gr.

Policy, has identified strategic energy technologies and designed roadmaps to tackle the individual barriers hindering their effective implementation across Europe. In the above context, a set of alternative pathways were designed by identifying factors that conform to the objectives of the EU and the SET-plan, in particular competitiveness and a clean, secure and efficient energy system. These pathways are positioned under two key uncertainties; the level of cooperation (i.e., cooperation versus entrenchment) and the level of decentralisation (i.e., decentralisation versus path dependency). The aim of this paper is to assess these pathways, through the active involvement of relevant stakeholders, according to their performance in key areas, such as the regulatory framework, market maturity, economic factors and stakeholder awareness towards a sustainable energy system, by using a multicriteria decision aid method (MCDA). Key challenge of our research is to assist European policy makers in drawing recommendations by exploring important elements, drivers and factors of the energy transition and the long term impacts of alternative mitigation options on the economy, the energy sector and technology development. The results could prove to be useful in further supporting strategic decision making in Europe's energy sector towards a clean, secure and efficient energy system.

Keywords: climate change, multicriteria decision aid, energy policy, Fuzzy PROMETHEE

INTRODUCTION

Nowadays, it is considered undeniable that innovation and climate challenges define the direction of a future EU energy system. In this context, Europe introduced the European Strategic Energy Technology Plan (SET-Plan), which aims to accelerate the development and deployment of low-carbon technologies with respect to the targets set for 2020 and 2050 (EC, 2009). The SET-Plan – as part of the technology pillar of European Energy and Climate Policy – has identified strategic energy technologies and designed roadmaps to tackle the individual barriers hindering their effective implementation across Europe. The 2013 Communication on Energy and Innovation has initiated a process of designing an integrated roadmap that identifies research and innovation challenges and needs of the EU energy system. But managing the transition to a more sustainable energy system has to take into account the societal, economic and environmental context in which the energy system is embedded. Therefore, a technology-oriented view has to be complemented with socio-economic research in order to increase our understanding of the complex energy system, improve our knowledge base for policy development and engage civil society in the transition of the energy system (Papapostolou et al., 2018; 2019).

In recent years, the global energy system has been marked by rapidly changing dominant trends and events, between the rapid expansion of shale oil and gas, the phase-out of nuclear energy, the drop in oil prices, and the internationally coordinated efforts to mitigate climate change since the November 2015 COP21 summit. This rapidly changing scenery in the energy sector significantly increases the need for decision-makers to understand the underlying interlinkages and implications of these emerging trends (Karakosta et al., 2015; Nikas et al., 2019). This rings particularly true for the European Energy Union and its efforts towards achieving the goals stated in the 2030 climate & energy framework (EC, 2014). An attempt to transition towards a decarbonised and fully-integrated single energy market necessarily looks several years, even decades ahead. It is therefore no surprise that a successful energy transition for Europe is contingent on understanding if and how emerging trends nowadays may actually be (weak) signals of forthcoming threats and opportunities (Karakosta, 2014). In order to successfully navigate towards a decarbonized future, it is more often than not necessary for decision-makers to rely on multicriteria decision aid methods to produce models that can estimate the long-term effects of individual policies and technologies on the energy system (Doukas et al., 2009; Karakosta and Psarras, 2012; Papapostolou et al., 2016; Karakosta, 2016; Papapostolou et al., 2017a; Papadogeorgos et al., 2017; Papapostolou et al., 2017b). To cope with the disparate preferences of decision makers, as well as to manage the uncertainty that arises when solving decision problems, a methodological assessment framework is developed based on an extension of the Preference Ranking Organization METHod for Enrichment of Evaluations (PROMETHEE) for group decision making. Making use of the popularity and suitability of Fuzzy PROMETHEE in managing energy sector problems (Andreopoulou et al., 2018; Nigussie et al., 2018; Nikas et al., 2018; Vujosevic and Popovic, 2016; Xenarios and Polatidis, 2015; Panagiotidou et al., 2015; Lerche and Geldermann, 2015), this study offers an original work able to shed light in the policy making problem related to sustainable energy transition.

ALTERNATIVES AND CRITERIA DEFINITION

The design of the transformation pathways is guided by a set of research questions, which are driver questions related on why the pathway scenario happens. The pathways adopt the widely-used 2 x 2 scenario typology to combine two main dimensions of uncertainty into four storylines spanning a wide possibility space. Figure 1 shows the scenario typology which varies two critical uncertainties: decentralisation vs. path dependency (x-axis); and cooperation vs. entrenchment (y-axis).

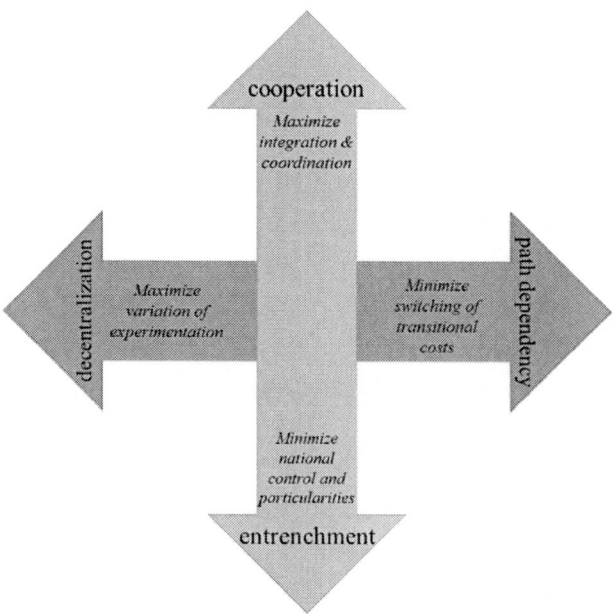

Figure 1. Proposed 2 x 2 scenario typology = cooperation-boundaries x decentralisation-path dependency.

These two dimensions of uncertainty create a possibility space which can be explored by the four contrasting storylines. Figure 2 shows the 2x2 variation in critical uncertainties facing the EU energy system, and how these uncertainties may combine to generate four different development pathways (Wilson and Kim, 2018). It is important that the interpretive detail of each storyline is internally consistent (avoiding tensions or contradictions), comprehensive (covers all relevant drivers and dynamics), and coherent (adds up to a meaningful whole). These are as follows:

Diversification. This pathway describes a decentralizing trajectory for the EU energy system in the context of cross-border cooperation and integration.

Directed Vision. This storyline describes a path-dependent trajectory for the EU energy system which is directed by the Commission's vision set out above for an ever-closer energy union.

National Champions. This pathway describes a path-dependent EU in which historical incumbency and national interests play a stronger guiding hand.

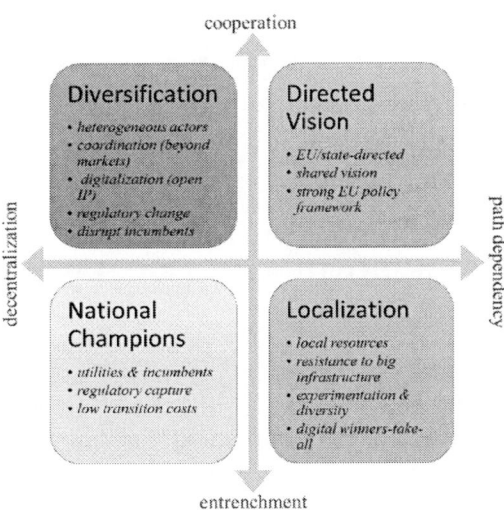

Figure 2. Transformation pathways at a glance.

Localisation. This storyline describes how the decentralising forces observable today in the EU start to chip away more forcefully at the centralised infrastructures, firms, and regulatory environments, but with marked national and local variation.

The nature of the problem is multidimensional, as it is of utmost importance for the policy makers to have the capability to assess the different pathways in a holistic approach, while taking into consideration implications in all various areas of influence. The criteria that were selected for the multicriteria assessment of the alternative pathways are:

- C_1 – Regulatory Framework. This criterion assesses the adequacy of the regulatory framework to support and ensure the implementation of the actions and policies proposed by each path.
- C_2 – Compatibility with Market. Given that Europe has a mature market, it is important to assess the extent to which each pathway is compatible with the current situation
- C_3 – Compliance with SET-Plan. This criterion reflects the extent to which each pathway achieves the goals of SET-Plan and evaluates its ease of implementation.
- C_4 – Stakeholder Awareness. This criterion assesses the extent to which those involved in each path are aware of climate change issues and are actively taking action to combat it.

METHODOLOGICAL FRAMEWORK

In the proposed methodological framework, the PROMETHEE method, developed by Brans (1982), is combined with fuzzy logic, introduced by Zadeh (1965), in order to exploit a method capable of tackling the transformation pathways problem.

The main methodological steps of the fuzzy PROMETHEE for group decision making that applied are the following:

- *Step 1:* Determine alternatives (m), evaluation criteria (k) and group of decision-makers (n).
- *Step 2:* Define linguistic variables, and their corresponding triangular fuzzy numbers, based on which the evaluation of the criteria's importance and the ratings of the alternatives will take place Chen and Hwang (1992).

Table 1. Linguistic variables and fuzzy numbers

Weights of criteria	Fuzzy Number	Ratings of alternatives
Very Low (VL)	(0.00, 0.00, 0.25)	Worst (W)
Low (L)	(0.00, 0.25, 0.50)	Poor (P)
Medium (M)	(0.25, 0.50, 0.75)	Fair (F)
High (H)	(0.50, 0.75, 1.00)	Good (G)
Very High (VH)	(0.75, 1.00, 1.00)	Best (B)

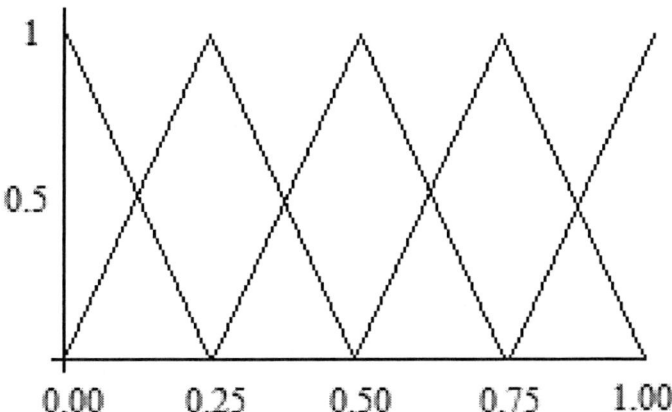

Figure 3. Membership function of triangular fuzzy numbers.

- *Step 3:* Aggregate decision-maker evaluations. A decision is derived by aggregating the fuzzy weights of criteria and fuzzy rating of alternatives from n decision-makers. Additionally, the preferences and opinions of n decision-makers with respect to j criterion (C_j) for the importance weight of each criterion and with respect to i alternative (A_i) for the rating of each alternative to each criterion can be calculated using the equations (1) and (2).

$$\widetilde{w_j} = \frac{1}{n}\left[\sum_{e=1}^{n} \widetilde{w_j^e}\right] = \frac{1}{n}\left[\widetilde{w_j^1}(+)\widetilde{w_j^2}(+)\cdots(+)\widetilde{w_j^n}\right] \quad (1)$$

$$\widetilde{x_{ij}} = \frac{1}{n}\left[\sum_{e=1}^{n} \widetilde{x_{ij}^e}\right] = \frac{1}{n}\left[\widetilde{x_{ij}^1}(+)\widetilde{x_{ij}^2}(+)\cdots(+)\widetilde{x_{ij}^n}\right] \quad (2)$$

The importance of decision-maker's i opinion is given by variable r_i and table R, as shown in equation (3), includes the importance weights of all decision-makers.

$$R = [r_1 \quad r_2 \quad \cdots \quad r_n] \quad (3)$$

Therefore, equations (2) and (3) are altered as follows:

$$\widetilde{w_j} = \left[\sum_{e=1}^{n} r_e \widetilde{w_j^e}\right] = \frac{1}{n}\left[r_1\widetilde{w_j^1}(+)r_2\widetilde{w_j^2}(+)\cdots(+)r_n\widetilde{w_j^n}\right] \quad (4)$$

$$\widetilde{x_{ij}} = \left[\sum_{e=1}^{n} r_e \widetilde{x_{ij}^e}\right] = \frac{1}{n}\left[r_1\widetilde{x_{ij}^1}(+)r_2\widetilde{x_{ij}^2}(+)\cdots(+)r_n\widetilde{x_{ij}^n}\right] \quad (5)$$

- *Step 4:* Construct a fuzzy decision matrix and compute the aggregated fuzzy weight of criterion.

$$\widetilde{D} = [\tilde{x}_{ij}]_{m \times k} = \begin{bmatrix} \tilde{x}_{11} & \tilde{x}_{12} & \cdots & \tilde{x}_{1k} \\ \tilde{x}_{21} & \tilde{x}_{22} & \cdots & \tilde{x}_{2k} \\ \vdots & \vdots & & \vdots \\ \tilde{x}_{m1} & \tilde{x}_{m2} & \cdots & \tilde{x}_{mk} \end{bmatrix} \quad (6)$$

$$\widetilde{W} = [\widetilde{w_1} \quad \widetilde{w_2} \quad \cdots \quad \widetilde{w_k}] \quad (7)$$

where $\widetilde{x_{ij}}$ is the rating of alternative A_i with respect to criterion C_j, and $\widetilde{w_j}$ is the importance weight of jth criterion.

- *Step 5:* Choose the Type of the preference function and determine the corresponding thresholds. The use of Type V (Linear preference function with indifference area) is considered to be more suitable (Figure 4) (Geldermann et al., 2000).

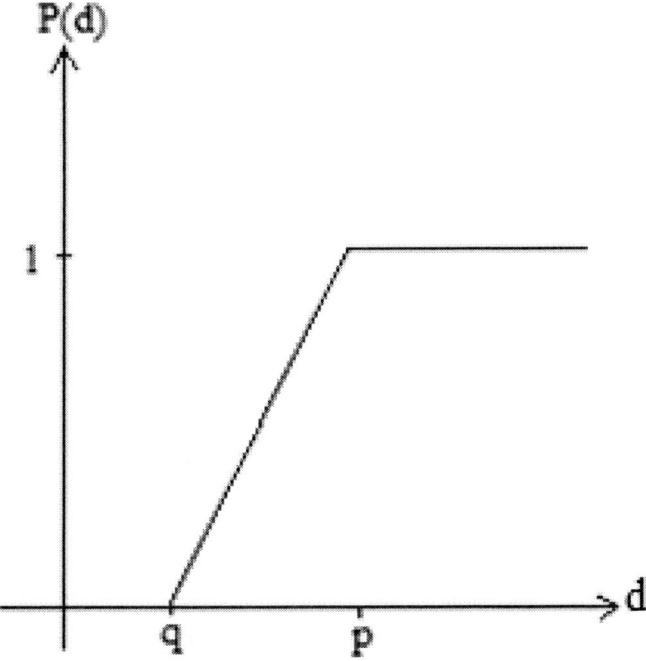

Figure 4. Type V preference function.

- *Step 6:* Generate the fuzzy multicriteria preference index $\tilde{\pi}$ for each pair of alternatives A_i, A_j according to equation (8).

$$\tilde{\pi}(A_i, A_j) = \sum_{t=1}^{k} \widetilde{w_t}(x) \, p_t(x_{it}(-)x_{jt}) \tag{8}$$

where $p_t(x_{it}(-)x_{jt})$ is the preference degree resulting from the comparison between alternatives A_i, A_j with respect to criterion t.

- *Step 7:* Calculate fuzzy Leaving Flow $\widetilde{\Phi}^+(A_i)$, as a measure of the superiority of the alternative A_i (9) and fuzzy Entering Flow $\widetilde{\Phi}^-(A_i)$, as a measure of the inferiority of alternative A_i (10). The difference of the aforementioned quantities produces the fuzzy Net Flow (11).

$$\widetilde{\Phi}^+(A_i) = \frac{1}{m-1} \cdot \sum_{j=1}^{m} \tilde{\pi}(A_i, A_j) \tag{9}$$

$$\widetilde{\Phi}^-(A_i) = \frac{1}{m-1} \cdot \sum_{j=1}^{m} \tilde{\pi}(A_j, A_i) \tag{10}$$

$$\widetilde{\Phi}(A_i) = \widetilde{\Phi}^+(A_i)(-)\widetilde{\Phi}^-(A_i) \tag{11}$$

Step 8: The final total ranking is achieved after defuzzification of fuzzy Net Flow (Geldermann et al., 2000; Ross, 2005; Nazari-Shirkouhi and Keramati, 2017). For the sake of brevity we define $\widetilde{\Phi}(A_i) = x_i$.

$$x_i^{defuzz} = \frac{\int x_i \cdot \mu(x_i)\, dx_i}{\int \mu(x_i)\, dx_i} \tag{12}$$

where $\mu(x_i)$ is the membership function of fuzzy number x_i.

The final ranking of the problem's alternatives is obtained directly from sorting all the defuzzified x_i.

It is worth mentioning that sensitivity analysis was conducted that helps to understand how the model variables react to input changes, and whether they are related to the data used to adapt the structure of the model, or to independent variables of the model (Wallace, 2000; Wainwright and Mulligan, 2005). Four scenarios were run as follows:

- 1st Scenario (reference): DMs are considered to be equal (33.33%) and their opinions contribute to the same extent in the final ranking.
- 2nd – 4th Scenarios: In these scenarios, greater emphasis is placed on the opinion of one DM each time (60%), while the views of the other two contribute secondarily to the final ranking (20% each).

RESULTS AND DISCUSSION

Based on the above-mentioned steps, this research used a total of three decision-makers (DMs) D_r, $r = \{1, 2, 3\}$, four different criteria C_j, $j = \{1, 2, 3, 4\}$, the four alternative pathways A_i, $i = \{1, 2, 3, 4\}$.

The next step concerns data collection so as to represent decision-makers' value system through definition of criteria weights and alternative ratings. Due to the diversity of stakeholders, three main decision-maker profiles were defined that represent: a policy maker (DM_1), an entrepreneur – representative of the energy industry (DM_2), and a researcher – representative of the academia (DM_3). Their views regarding the importance of each criterion and the evaluation of the alternatives are shown in Tables 2 and 3.

The results of the method application for a group of three decision makers are the following:

Based on the Fuzzy PROMETHEE ranking, Directed Vision has outranked all other pathways which means that it faces less risks of failure and has better prospects when implemented. Second best in the ranking comes the National Champions mainly because of its relatively worse performance in the Compliance with SET-Plan criterion (compared to the Directed Vision). The other two pathways have negative flows, which means that they perform worse in almost every aspect. It is worth mentioning that even when each scenario is assessed individually National Champions and Directed Vision are found on top of all three rankings, reinforcing the conclusion that these two pathways are commonly accepted as the safer approaches for the energy future.

Table 2. Fuzzy PROMETHE ranking

Alternatives	Scenario			
	1st	2nd	3rd	4th
Diversification	4	4	4	4
Directed Vision	1	1	2	1
Localization	3	3	3	3
National Champions	2	2	1	2

CONCLUSION

The choice of method proves to fit the needs of the specific problem since it enables the DMs form a holistic view which encompasses all the individual, and controversial views regarding each pathway. DMs perspectives were successfully combined with the pathways' storylines to generate qualitative insights as input into more detailed energy systems modelling and analysis. Thanks to the use of linguistic variables, the process becomes friendlier to the DMs and at the same time the uncertainty in their words is taken into consideration and it is encoded in the input data. In addition, the combination of different.

As further research, more scenarios could be produced based on the opinions of different decision-makers, while more decision makers could be engaged from several target groups (Society, NGO, government, etc.). Additional evaluation (qualitative and quantitative) criteria could be also used in order to capture more important aspect of the energy system decarbonisation, e.g., financial and societal impact, while a different perspective could be adopted, so that the pathways are assessed and compared based on their performance to the prevailing technological aspects (e.g., case studies on renewable energy, smart grid, energy efficiency, sustainable transport, CCS and nuclear safety).

ACKNOWLEDGMENTS

We gratefully acknowledge support from the EC, grant 691843, SET-Nav - Navigating the Roadmap for Clean, Secure and Efficient Energy Innovation (www.set-nav.eu), for the valuable contribution with regard to identifying alternative transformation pathways for the European energy system, which are presented in deliverable 'Mapping empirical analysis of the EU's energy innovation system onto storylines of future change', April 2018. Their work in this respect was invaluable for this research. The contents of the paper are the sole responsibility of its authors and do not necessarily reflect the views of the EC.

REFERENCES

Andreopoulou, Zacharoula, Koliouska, Christiana, Galariotis, Emilios, and Zopounidis, Constantin. "Renewable energy sources: Using PROMETHEE II for ranking websites to support market opportunities." *Technological Forecasting and Social Change* 131 (2018). 31-37. doi: https://doi.org/10.1016/j.techfore.2017.06.007.

Brans, Jean P. "L'ingenierie de la decision, l'laboration d'instruments d'aidea la decision. Colloque sur l'Aidea la Decision." Faculte des Sciences de l'Administration, Universite Laval (1982). ["The engineering of the decision, the development of instruments to help the decision. *Symposium on the Aid to Decision.* "Faculty of Sciences of the Administration, Laval University (1982).]

Chen, Shu-Jen, and Ching-Lai Hwang. "Fuzzy multiple attribute decision making methods." *Fuzzy multiple attribute decision making.* Springer, Berlin, Heidelberg, 1992. 289-486.

Doukas, Haris, Charikleia Karakosta, and John Psarras. "A linguistic TOPSIS model to evaluate the sustainability of renewable energy options." *International Journal of Global Energy Issues* 32.1-2 (2009). 102-118.

EC - European Commission. Communication from the Commission to the European Parliament, the Council, the European Economic and Social Committee and the Committee of the regions. *A policy framework for climate and energy in the period from 2020 to 2030*, Communication, COM(2014) 015 final. Brussels: European Commission.

EC - European Commission. *Investing in the Development of Low Carbon Technologies* (SET-Plan). COM(2009) 519.

Geldermann, Jutta, Thomas Spengler, and Otto Rentz. "Fuzzy outranking for environmental assessment. Case study: iron and steel making industry." *Fuzzy sets and systems* 115.1 (2000). 45-65.

Karakosta, Charikleia, and John Psarras. "Fuzzy TOPSIS approach for understanding a country's development priorities within the scope of climate technology transfer. Adv." *Energy Res.* 9 (2012). 123-149.

Karakosta, Charikleia, Marinakis, Vangelis, Papapostolou, Aikaterini, and Psarras, John. "Benefits and Costs Sharing through RES Electricity Cooperation Between Europe and Third Countries." *Paper presented at the 3rd International Conference on Energy Systems and Technologies - ICEST 2015, Cairo, Egypt, February 16-19, 2015.*

Karakosta, Charikleia. "A Holistic Approach for Addressing the Issue of Effective Technology Transfer in the Frame of Climate Change." *Energies* 9.7 (2016). 503.

Karakosta, Charikleia. "Technology transfer for effective policy design. APRAISE (Assessment of Policy Interrelationships and Impacts on Sustainability in Europe)" *Paper presented at the Summer School, University of Ljubljana, Faculty of Electrical Engineering*, Ljubljana, Slovenia, 25-29 August 2014.

Lerche, Nils, and Jutta Geldermann. "Integration of prospect theory into PROMETHEE-a case study concerning sustainable bioenergy concepts." *International Journal of Multicriteria Decision Making* 5.4 (2015). 309-333.

Nazari-Shirkouhi, Salman, and Abbas Keramati. "Modeling customer satisfaction with new product design using a flexible fuzzy regression-data envelopment analysis algorithm." *Applied Mathematical Modelling* 50 (2017). 755-771.

Neofytou, Hera, Karakosta, Charikleia, Caldés Gómez, Natalia. "Impact Assessment of Climate and Energy Policy Scenarios: A Multi-criteria Approach" In *Understanding Risks and Uncertainties in Energy and Climate Policy*, edited by Doukas Haris, Flamos Alexandros, Lieu Jenny. 123-142 (eds) Springer, Cham, 2019 ISBN 978-3-030-03151-0.

Nikas, Alexandros, Gkonis, Nikolaos, Forouli, Aikaterini, Siskos, Eleftherios, Arsenopoulos, Apostolos, Papapostolou, Aikaterini, Kanellou, Eleni, Karakosta, Charikleia, and Doukas, Haris. "Greece: From near-term solutions to long-term pathways: risks and uncertainties associated with the national energy efficiency framework." In

Narratives of low carbon transitions: Understanding risks and uncertainties, edited by Susanne Hanger-Kopp, Jenny Lieu, Alexandros Nikas. 180-197 Series: Routledge Studies in Energy Transitions, 2019. ISBN 9781138311589. doi: https://doi.org/10. 4324/9780429458781.

Nikas, Alexandros, Haris Doukas, and Luis Martínez López. "A group decision making tool for assessing climate policy risks against multiple criteria." *Heliyon* 4.3 (2018). e00588.

Papadogeorgos, Ioannis, Papapostolou, Aikaterini, Karakosta, Charikleia, and Doukas, Haris. "Multicriteria Assessment of Alternative Policy Scenarios for Achieving EU RES Target by 2030." *Strategic Innovative Marketing*. Springer, Cham, 2017. 405-412.

Papapostolou, Aikaterini, Charikleia, Karakosta, and Haris Doukas. "Analysis of policy scenarios for achieving renewable energy sources targets: A fuzzy TOPSIS approach." *Energy & Environment* 28.1-2 (2017a). 88-109.

Papapostolou, Aikaterini, Karakosta, Charikleia, and Doukas, Haris *"Assessment of Alternative Policy Strategies towards a Decarbonised Energy System: A Fuzzy-PROMETHEE Approach,"* Paper presented at the ICEF IV, 4th International Conference on Engineering and Formal Sciences, Amsterdam, The Netherlands, 14-15 December 2018.

Papapostolou, Aikaterini, Karakosta, Charikleia, Doukas, Haris, and Psarras John (2019). *"Assessment of Alternative Scenarios for the Decarbonisation of the Energy System,"* Paper presented at the 6th Student Conference of HELORS, Xanthi, Greece, 28 February - 2 March 2019 (in Greek).

Papapostolou, Aikaterini, Karakosta, Charikleia, Marinakis, Vangelis, and Alexandros, Flamos. "Assessment of RES cooperation framework between the EU and North Africa: a multicriteria approach based on UTASTAR." *International Journal of Energy Sector Management* 10.3 (2016). 402-426. doi: 10.1108/IJESM-12-2014-0007.

Papapostolou, Aikaterini, Karakosta, Charikleia, Nikas, Alexandros, Psarras, John. "Exploring opportunities and risks for RES-E deployment under Cooperation Mechanisms between EU and Western Balkans: a multi-criteria assessment." *Renewable and Sustainable Energy Reviews* 80 (2017b): 519-530. doi: https://doi.org/10.1016/ j.rser.2017.05.190.

Ross, Timothy J. *Fuzzy logic with engineering applications.* John Wiley & Sons, 2005.

Vujosevic, Milica L., and Milena J. Popovic. "The comparison of the energy performance of hotel buildings using PROMETHEE decision-making method." *Thermal Science* 20.1 (2016). 197-208.

Wainwright, John, and Mark Mulligan, eds. *Environmental modelling: finding simplicity in complexity.* John Wiley & Sons, 2005.

Wallace, Stein W. "Decision making under uncertainty: Is sensitivity analysis of any use?." *Operations Research* 48.1 (2000). 20-25.

Wilson, Charlie, and Kim, Yeong Jae. *"Mapping empirical analysis of the EU's energy innovation system onto storylines of future change."* Deliverable of the "SET-Nav - Navigating the Roadmap for Clean, Secure and Efficient Energy Innovation" project (2018). http://set-nav.eu/

Xenarios, Stefanos, and Heracles Polatidis. "Alleviating climate change impacts in rural Bangladesh: a PROMETHEE outranking-based approach for prioritizing agricultural interventions." *Environment, development and sustainability* 17.5 (2015). 963-985.

Zadeh, Lotfi A. "Fuzzy sets." *Information and control* 8.3 (1965): 338-353.

In: A Comprehensive Guide to Energy ... ISBN: 978-1-53616-728-3
Editor: Shannon Alvarado © 2020 Nova Science Publishers, Inc.

Chapter 3

ELECTRICITY TRANSITIONS IN GERMANY: THE TRANSFORMATION OF A STRATEGIC ACTION FIELD

*Gerhard Fuchs**

Institute of Social Sciences, University of Stuttgart,
Stuttgart, Germany

ABSTRACT

The chapter draws on recent developments in the field of electricity generation and distribution in Germany. We analyze decentralized forms of electricity generation and distribution. Pioneers of this development seized opportunities connected with broad institutional changes to discredit the status quo and work out legitimations for their new model of how to generate and distribute electricity. Our analysis suggests important differences in how actors legitimate novel organizational forms in emerging vs. mature fields and it underscores the need for theories of social change that explicitly account for field context.

---

* Corresponding Author's E-mail: gerhard.fuchs@sowi.uni-stuttgart.de.

Keywords: legitimacy, action fields, governance, electricity transition, Germany, renewable energies

1. INTRODUCTION

In order to fight climate change and its dire consequences an energy transition is required. At least this is the consensus among climate scientists, large chunks of the population in developed countries, some governments and of course the vast majority of social scientists. But, as it has been repeatedly stated, the overall energy transition is lagging behind even modest time tables, and countries like Germany, which for some time have been considered to be forerunners, are now scrapping their goals in favor of conventional economic growth and the preservation of employment in threatened traditional industries.

Many conflicting interests are obviously at work and looking at the possible realms of decisions to be made, we are dealing with a number of different arenas from government, business up to private households. Some things definitely have to be dealt with by government decisions, but even the government usually is not one compact entity equipped with sufficient steering capacities. The relevant actor spectrum and the actor constellations differ depending on the specific area, time and problems we are dealing with.

Thus, there is no "one" governance of the electricity transition. Governance problems and problems of legitimation have to be considered case-wise. Practices are situated in specific environments, which put different, sometimes conflicting demands on the actors. Usually – that is in stable fields[1] - the status quo is considered to be legitimate by the people engaged in a specific practice (Meyer and Rowan (1977)). Practices, however, can be delegitimized due to changes in the environment. The

---

[1] When using the term "field," I do this in the tradition of field theory as summarized by Martin (2003). Fields are meso-level social orders where actors (who can be individual or collective) interact with knowledge of one another under a set of common understandings about the purposes of (in our case) a specific sector, a field, the relationships there (including who has power and why), and the sectors' rules.

practice of generating electricity using coal is not being delegitimized by the dominant actors in the field, but by an environment, which considers the burning of coal to be damaging to the health of people and the environment. As Lawrence and Philipps (2004: 705) have argued: "*The delegitimating effects of change in macro-cultural discourses may deinstitutionalize the activities of a field despite the actions of local actors.*" Legitimation has furthermore to be seen in relation to the relevant level of decision-making. The legitimation of decisions made by the federal government and/or the parliament works different from the legitimation of the implementation of a decision in a specific place and at a specific time. It is an often-discussed issue that people in principle are in favor of renewable energy (RE) installations, but they might protest against the construction of a wind turbine in their immediate neighborhood. Different principles of legitimacy are at work here.

In our case, the German system of electricity production and distribution has undoubtedly undergone a thorough process of institutional transformation and changing legitimations – unlike other components of the energy system. A characteristic element of such a process is that an ultimate outcome cannot be predetermined. It is also unlikely that a big transition is guided by some unitary mechanism or guiding principle – especially true for the case of the German electricity system.

On an empirical level, we will be asking how the transformation of the German system was legitimated and realized against the resistance of incumbent actors and a wavering government. On a conceptual level, the article will try to contribute to the ongoing discussion in the social sciences concerned more with the way processes (Abbott 2016) and structures (Martin 2009) develop.

We will proceed as follows. Section two will present the conceptual framework of the paper. Since the research on which this manuscript is based cannot be described in detail due to space constraints, a short section three on methods will follow. Section four will reconstruct the process of change in the German system of electricity generation and distribution. A conclusion will summarize the arguments.

2. Fields, Practices and Actors

It seems to be generally acknowledged that in the past sociology was primarily concerned with static structures and regularities, i.e., social facts (Martin 2009). The problem of change, on the other hand, was given short thrift. As such, we are in a sense still at the beginning of introducing change into sociological theory. Abbott (2016), for example, is concerned with the principles of a processual sociology; Padgett and Powell (2012) investigate the origin and development of new organizations and markets; Fligstein and McAdam (2011, 2012) are developing a field theory combining aspects of stability and change in social relations. All these authors share the idea that structures and institutions should not be treated as something given and, as such, constraining, but as something that comes into existence, develops, and eventually might dissolve again. Martin (2009) makes it clear that such an approach does not mean turning sociology into history. If we look upon structures and institutions as processes, we will naturally be treating them as being transient, coming from somewhere, having a certain configuration in the present, which will be different in the future. We will, however, look for generic principles that tell us how structures are formed and stabilized and how they may eventually be destroyed again. This turns the attention of the researcher away from a traditional concern with "why" questions, putting mechanisms and "how" questions in the foreground (White 2008).

It can be assumed that a stable social practice revolves around a common understanding of what is going on in a field. There are actors who wield more or less power in influencing the practice; they can be called incumbent and challenger actors, respectively. There is a common set of rules, defining legitimate actions and a broad interpretative framework that individual and collective actors share to make sense of what others within the field are doing. This all happens within an environment that is constantly changing and posing new challenges to the participating actors. Political actors play an eminent part in this environment by setting the regulatory framework. Changes in practices are advanced by challenger actors coming either from outside the established fields or from disadvantaged "niches" within the field. In order to be able to change an established practice, challengers need to be able to take advantage of changes in the surrounding environment that—properly framed – will help them to delegitimize existing practices. Learning is done along the lines of improving an established practice. Developing new practices or changing them substantially involves unlearning, which is promoted by the challengers and potentially forced on the incumbents. Such a process of unlearning can be conceived as disruptive innovation. Slightly modifying the stance of Padgett and Powell (2012), who distinguish between innovation and invention, one can say: *incremental innovations* improve on existing ways of doing things, while *radical innovations* change the ways things are done. Based on this it can be determined that an incremental innovation in the field of electricity generation aims at improving the existing ways and procedures (fossil fuel- and nuclear power-based electricity generation by established actors) and at gradually adapting to new demands coming from the environment (more RE installations). Radical innovation is concerned with new ways of generating electricity, engages new actors, and builds new networks.

Analyses of processes of field transition have shown that such processes as well as their outcomes are difficult to predict and might take different forms such as: (a) a re-imposition of the old regime with some adjustments (adaptation); (b) the break down into unorganized social space; (c) the partitioning into several spaces (e.g., renewable vs. traditional energy generation) (differentiation); (d) the development of a wholly new regime (cf. Mahoney and Thelen 2010, Fligstein and McAdam 2011). The term "transformation" will be reserved for the last option.

3. A BRIEF WORD ON METHODS AND METHODOLOGY

The reported research for one is based on a historiographic approach: documents, archival materials, texts, web materials and presentations were used to develop a historical account of the electricity transition (Ventresca and Mohr 2002). In addition, expert interviews were conducted and a close cooperation with mainly challenger actors was achieved. The resulting narrative account is intended to sharpen, illustrate, and ground the arguments, not to provide an empirical test or a comprehensive history. Using historical data allows to trace a process that occurred over several decades, thereby overcoming the temporal limitations of ethnographic and other field methods (Jones and Khanna 2006).

The electricity transition in Germany provides an excellent context to study field emergence and processes of delegitimation. A variety of new organizational forms were developed by challenger actors dissatisfied with the status quo and eager to change the dominating system. Framings to delegitimize the dominating system based on fossil fuels and nuclear energy and operated by big energy providers in close partnership with big politics were developed as well as schemes for a "new" system of electricity generation and distribution. The old regime meanwhile has gone overboard, some of the demands of the challenger actors have been fulfilled, the future of the old utilities as of yet is undecided. The retracing of the history of the German electricity transition showed that the institutional entrepreneurs in this context - the "agents of legitimacy" (Dacin et al. 2002, 47) - were not a (small) group of identifiable persons of national prominence, who spread an identical message. The initiative was usually with locally embedded persons, which did not have a standing in the field proper, but for different reasons were trying to delegitimize the rules of the mature field.

4. INSTITUTIONAL TRANSFORMATION

Analyzing processes such as electricity transitions makes one think about time periods. Determining periods or distinguishing between different phases of development in a process has always included some element of arbitrariness. This is indicated by questions such as where to start and where to end, and how to avoid the impression that the present state of affairs is to be considered the logical outcome of a development full of contingent events. If a study is written in 2019 rather than in 2017, the process results might be quite different, and it would be ludicrous to assume that the logical conclusion of the activities of the year 2017 will necessarily be what comes into existence in 2019. Clearly if one wants to retrace a process, the elements highlighted need to be connected to one another, in order to show the flow, but as mentioned before one should not suggest that the flow has been without alternatives (Abbott 2001).

A relational field approach fully incorporates at least two types of actors or agencies occupying different positions within the social space or field and bound together in a relationship of mutual dependence or struggle. In this way, at the beginning of the electricity transition process it can be distinguished between the incumbent actors, dominating the field, and the challengers, at the fringes or even outside of the field, who are eager to disturb the dominant field practices. The roots of the dominant system of electricity generation and distribution in Germany in the 1990s differ from those of the RE movement, which can be considered as the challengers. Insofar there are two processes running parallel for some time. They have one common reference point: the discussion about the legitimacy of nuclear energy. The dominant industrial and political actors for a long time considered nuclear energy to be an important building block of the German electricity system. A social movement against nuclear energy, however, led to a stop of construction plans, and especially after the Chernobyl accident the antinuclear sentiment won broad public support and attempted to delegitimize nuclear energy. The attribution of a threat to nuclear energy and later on to climate change was successfully achieved by the challengers. Actors associated with the threat were the big energy companies and parts of the government(s). The anti-nuclear power movement had demonstrated that mobilizing can be effective and successful. In their efforts to mobilize, RE actors were also framing a collective attribution of opportunity, by developing the idea that RE and a new decentralized system of energy generation and distribution would help in coping with the threats to the environment. This attribution is a characteristic element, which has also been found in other researched cases in a wide array of fields: "The legitimacy of a new organizational form in an emerging field is increased when institutional entrepreneurs identify and promulgate contradictions between the status quo and broad institutional logics external to their field and theorize their organizations as solutions to these problems" (David/Sine/Haveman 2013: 365).

The success of the movement and the specific German way of dealing with problems (e.g., as opposed to the UK or France) depended on specific changes in the environment of the movement. Thirty years of social movement research have affirmed the important role that favorable "political opportunities" play in shaping the emergence and development of contentious politics (Kriesi 2004). We will try to trace the development of the German electricity transition along the lines of changes in political power constellations. The changes in power constellations, of course, do not immediately or necessarily translate into changes in what is considered a legitimate social practice. Nevertheless, they influence opportunity structures in the sense that the changed composition of the government did go along with changing attitudes towards RE deployment.

4.1. Phase One (–1998): Building Niches and Developing Markets: Questioning the Legitimacy of Established Practices

In the 1990s, there was a straightforward consensus regarding who acts legitimately and how legitimate activities in the field of electricity generation and distribution are to be conducted. The dominant idea was that electricity generation has to rely on fossil fuels and nuclear energy. Electricity is to be produced in big power plants, owned by actors able to make capital-intensive long-term investments. Electricity, generated in some distance from the points of consumption is transported via big cross-country grids to the end user. Before electricity markets were formally liberalized in 1998, the German electricity sector was organized on the basis of a "natural monopoly." The market was shaped by nine vertically integrated public utilities, which provided electricity within specific territorially fixed boundaries (Bontrup and Marquardt 2010, 20f). The nine companies mainly served as the backbone for 80 regional supply companies and about 900 municipal utilities. Most of the municipal utilities held little or no production capacity at all and concentrated on matters of distribution. This setup reflected the decentralized way the electricity system in Germany had developed historically and the typical quasi-corporatist organizational setup for public services to be found elsewhere as well (e.g., post, telecommunications, and social insurance).

Except for some large industrial companies who owned power plants to help operate their production units, independent energy producers were generally considered to be illegitimate. Owners of decentralized RE systems did not get permission to connect their installations to the grid. The regulatory basis for this was the German Energy Act (EnWG) of 1935. The law guaranteed the supply monopoly for supposedly economic reasons. It was primarily justified by the argument that regional monopolies would enable the production of power at the lowest costs. For the dominant actors in this period, learning meant extending the reach of their business models to the former German Democratic Republic (GDR) and preparing for liberalization. In their optimization efforts, they were not searching for new technological or organizational options.[2] Anyway, why would they consider it necessary to abandon a successful legitimate practice? They were busy with further developing their privileged position in the field.

The search for new options did take place elsewhere: in the RE movement. The origins of this movement can be found mainly in the anti-nuclear power movement and to a lesser extent in protest movements against big industry/big government projects. Some members of the movement were scientists who were eager to develop alternatives to nuclear energy; many others were (organized) activists, who were willing to build experimental installations. An important window of opportunity for these groups opened following the rising electoral successes of the Green party in the 1980s and 1990s, which eagerly picked up the energy issue. In some communities where the green movement was strong, experiments were conducted with local energy solutions, operating with different organizational and technological mixes. There was little direct competition between incumbents and challengers, which were not yet considered to be relevant. The incumbents, however, did lobby politicians, obstruct RE programs, and use the courts to stop RE development.

---

[2] On the differentiation between optimizing and searching (as a characteristic of innovative firms), see Lazonick (2005).

The challengers emphasized their moral position towards nuclear energy and the climate crisis and thus emphasized altruism. They deemphasized their self-interests and highlighted instead the benefits of their activities for the overall society.

Political decision makers at the federal level were mainly concerned with reorganizing the electricity system in the former GDR and preparing for liberalization, an issue especially advanced by the European Commission. Government attitudes towards RE were reluctant and haphazard. Less by persuasion than as a sign of being open to the public mood against nuclear energy, technology-based support programs for small wind turbines were first introduced in the late 1980s (Neukirch 2010), which were later extended to solar energy; both were funded by the ministry for education and research (as opposed to the ministry of economic affairs, which was concerned with liberalization, market framework, and nuclear energy). The opponents of RE considered the programs to be too small and unimportant to be relevant and actually believed that it would demonstrate that RE is not viable in Germany. They therefore let it pass. For the supporters of RE, it was a major success, a symbol for successful collective action. Now the independent operators of turbines got the guarantee that utilities not only had to connect their installations to the general grid, but that they were also obliged to pay fixed tariffs for the energy supplied.

In the mid 1990s, however, there was general insecurity about the future stance of the government towards RE. The then minister of environmental affairs, Angela Merkel, proclaimed in the mid 1990s that it is not conceivable that RE will play any significant part in the future German electricity mix. Nevertheless, the 1990s proved to be a period in which RE actors could experiment with various designs, establish organizations, build up public support for RE, and extend the advocacy coalition. Using the language of transition theory, one could say that RE activities in this period were (not necessarily protected) niche activities, while coalition formation took place on a much broader scale.

4.2. Phase Two (1998–2008): The Politics of Layering: Developing Competing Legitimations

The second phase from 1998 to 2008 is dominated by the effects of the formation of a new federal government and the dynamic growth, which both incumbent actors and challenging niche actors were experiencing. The main disputes in this phase were concerned with the issue of the legitimacy of the different energy sources (coal, nuclear, renewables). In our case, this meant that the interpretation of what is a legitimate action became layered depending on which subfield it was taking place in. Yet there was no impression of a general crisis. Let us look at the details.

The year 1998 featured a change in government from a conservative one to one controlled by a coalition of the Social Democratic Party (SPD) and the environmentally focused Green Party (Bündnis 90/Die Grüne). Under their leadership, important but also slightly conflicting regulatory decisions were passed. Following an EU market directive aiming at liberalizing the European energy markets, an amendment to the Energy Economy Law (EnWG) was put into effect in 1998. Liberalization supported strong concentration in the market. A wave of mergers changed the playing field from one in which a number of vertically integrated energy companies were operating to one in which the "Big 4" (RWE, E.ON, EnBW, and Vattenfall) were loosely competing against one another. The Big 4 acquired shares in smaller regional suppliers as well as municipal utilities. The latter were affected by the competition with private energy suppliers, which resulted in considerably lower profits, in turn affecting the municipal budgets and the capacity to subsidize, for example, public transport. Hence, many of them were swallowed by the Big 4 or began to refrain from producing energy. The market share of the Big 4 subsequently grew to 82% of the electricity production capacity and 90% of electricity generated in 2003 and 2004 (Bundesnetzagentur 2007, 60).

With respect to RE important regulatory changes happened, which eventually contributed to far-reaching changes in the balance of power in the field. In 2000, the Renewable Energies Law (EEG) was passed. It provided incentives for investments in renewable energy generation by obliging grid operators to give priority to connecting such facilities. It also guaranteed a consistent minimum payment for the electricity produced for a period of 20 years and thus ensured investment and planning security. The level of financial support was held flexible depending on the maturity of the different technologies. This especially included much higher remuneration for photovoltaic installations, making this technology economically feasible for the first time (Hoppmann, Huenteler, and Girod 2014). In the end, the EEG helped boost the expansion of renewable energy from 6.6% of the electricity generated in 2000 to 14.5% in 2008.[3] For our purposes, it is equally important to note that the specification of the actors who can legitimately produce electricity in this expansion had changed significantly. Now new types of actors, i.e., ones in addition to the established utilities, were allowed to generate electricity and connect to the grid. Consequently, new types of legitimate social practices evolved beyond certain limited niches.

The federal law was inspired by local organizational efforts, which had used feed-in tariffs introduced over the 1990s in cities like Aachen, Freising, or Hammelburg, which provided legitimacy and ammunition for federal policy entrepreneurs like the SPD politician Hermann Scheer (Fuchs and Wassermann 2012). It was additionally supported by the lobbying activities of a broad advocacy coalition – consisting of NGOs like Eurosolar, Förderverein Solarenergie, and Greenpeace, but also industrial associations and even one incumbent utility (Preussen Elektra) – which intensified pressure on federal policymakers (Jacobsson and Lauber 2006, 267). Besides ecological arguments, this coalition pointed to the economic benefits of RE. They highlighted the first mover advantages of an early adoption of renewable energy and the benefits for the national economy from the growth of the RE industries. Furthermore, they claimed that the electricity market was distorted, since the prices for electricity from fossil sources did not include the external costs resulting from environmental pollution and climate change. Because of this, renewables would be much cheaper than its opponents were claiming (Hirschl 2008, 193). As the renewables industries grew in size and professionalization (e.g.,

---

[3] Data from the AG Energiebilanzen (https://www.ag-energiebilanzen.de/).

photovoltaic module producers, wind-turbine manufacturers) as well as due to the increasing contribution to regional value creation (especially in eastern Germany), there was growing support from previously critical parts of the SPD and the CDU (Lauber and Jacobsson 2016). The legitimation strategies now emphasized clearly in equal manner both the economic and the ecological benefits of RE.

In parallel, it was decided to phase out nuclear power plants. Following the *Atomkonsens* (consensus agreement on atomic energy) of 2000, the Atomic Energy Act was amended in 2002. The amendment stipulated an end to the construction of new nuclear power plants and limited the running time of the existing ones. The incumbents in the following changed their legitimation strategies. They stressed the security of supply guaranteed by conventional power plants and nuclear power, and the importance of the latter for climate change mitigation. They highlighted the benefits for the national economy and the opportunity to finance the switch to a renewable energy future by utilizing the profits made from nuclear energy. They also claimed that nuclear power plants would be much more flexible than generally assumed and would thus be perfect partners for renewable energy.

The organizations representing the vision of a decentralized energy supply developed strongly and successfully during these years. Mautz, Byzio, and Rosenbaum observed that the professionalization, stabilization, and differentiation of decentralized actors and diffusion networks increased and the general "social opening" of the electricity sector proceeded further – although the developments differed for the various types of technology (Mautz, Byzio, and Rosenbaum 2008, 83). Mautz and Rosenbaum differentiate four main types of actors involved in electricity generation in this period: homeowners, citizen action groups, private investors, and farmers. For all the renewable technologies, the multiplier function of civil society actors was important. The activities in the RE sector were thus still being driven forward by decentralized and newly organized actors.

The developments in phase two led to two different fields within electricity generation ("layering" or differentiation, see above). On the one hand there was the liberalized electricity market, dominated by the big four utilities relying on fossil fuels and nuclear energy. The new market environment allowed them to grow significantly by swallowing smaller competitors and realizing benefits of scale. On the other hand, there was the RE sector, which was regulated by different principles and which was populated by different actors. Here we found mainly small entrepreneurs, individual citizens, farmers, and cooperatives, who unlike the actors in the traditional field were not looking to optimize their business strategies, but who were struggling to build a new market framework that would allow them to act profitably. The motivating forces for many of these actors, however, were not primarily economic motives, but the intention to change the electricity system.

The learning trajectories of the two actor groups were thus very different. Direct competition between the two groups did not take place. Especially the RE group did not have significant problems from internal competition. The market for RE was developing quickly, and new actors were coming in on a continuous basis. As Lawrence and Philipps (2004, 706) have argued, institutional entrepreneurship in emerging fields is likely to be associated with rapid imitation and relatively little conflict. The immediate effects of institutional entrepreneurship in emerging fields are likely to be highly uncertain and therefore the strategies of institutional entrepreneurs are more likely to be emergent than intended. Unlike in the old field, hierarchies were not present and no small number of organizations was in a position to dominate the fragmented field.

Even without direct economic competition, the incumbents continued their opposition to RE. They used two main delegitimation strategies. (1) They claimed that German energy policy disregarded economic issues in a one-sided manner and would predominantly favor ecological concerns. They painted a gloomy picture of a coming deindustrialization and a shrinking competitiveness of the German economy. (2) They engaged in direct lobbying activities to get rid of the EEG or at least to alter it to their advantage. These lobbying activities were strengthened in the aftermath of the end of the red-green government in 2005.[4]

---

[4] For a more detailed analysis see Hirschl (2008): 169, 195 and Suck (2008).

4.3. Phase Three (2009–2013): Transformation and Disruptive Innovation: The Break Down of Traditional Legitimation Strategies

Important regulatory changes were again introduced in phase three. Phase three, however, was dominated by global developments external to the field, which eventually led to a significant crisis in the field. The success of RE had enticed many other countries to develop national industries. This was especially true for China, which began to produce cheap equipment and export it aggressively. The ensuing price competition proved detrimental to many German producers. This, however, affected not only the producers, but also the RE advocacy coalition. As mentioned previously, the actors who had supported RE for economic reasons, based especially on structural policy arguments, began to lose interest, especially since a newly formed federal government (2009) made it clear that it would not support the ailing industry. The big utilities also ran into problems. Internal growth in Germany was limited by competition laws and by a growing RE contribution to the electricity mix. The internationalization of their business proved to be largely a failure as a consequence of the global finance crisis. Growth prospects were thus dim. Finally, the Fukushima catastrophe had an important and decisive impact on electricity-related issues. Old and new actors were now forced into direct competition. With RE leaving their niche, formerly hidden conflicts and parallel developments in a differentiated field became increasingly apparent and obvious once the technology mix began to be shaped more and more strongly by high shares of RE, which especially at peak times increasingly displaced conventional energy sources in the electricity mix. The incumbents thus had to face a direct attack on their market position. This was the end of layering and established business practices.

This phase of the transformation was again characterized by contradictory political and regulatory developments, unstable coalitions among different actors in the field, and new conflicts and struggles. The new insecurity and contradictions became obvious in 2010. On the one hand, the government amended the Atomic Energy Act and extended the lifetime of nuclear power plants. Hence, for a short period it seemed as if the incumbent energy providers had scored an important victory against the challengers and relegitimized nuclear energy. One year later, however, the nuclear power coalition faced a clear backlash: in the aftermath of the Fukushima accident, the government reinvigorated the nuclear phase out and the so-called Energy Transition decision was passed, which made the general shift to RE official government policy. The political decision reflected the general mind-set of the German population. There had always been a strong opposition to nuclear power (Europäische Kommission 2008), and this rejection increased substantially after the Fukushima accident. Furthermore, the population explicitly supported the German energy transition and especially the idea of building a more decentralized infrastructure and distributed power generation (TNS Emnid 2013). The population mistrusted the big power companies and was explicitly in favor of RE (Scheer, Wassermann, and Scheel 2012).

On the other hand, the political coalition supporting RE began to crumble. The German RE industry suffered severely from cheap, mainly Chinese, imports. The bulk of the industry disappeared again, and along with it the hopes of local and state politicians for economic rewards. The government was not providing assistance but was rather supporting a process, which it called "market adaptation." The farmers' association began to lobby against RE because it was afraid that the economic success of energy farmers would undermine their main business of lobbying for subsidies for regular farmers. The old political coalition of red-green supporters of RE was excluded from power and even within the SPD the supporters of RE lost influence as concern over job losses in the heavily unionized old utilities grew. This all led to regulatory amendments, which generally tried to make life for RE more difficult, especially for small, decentralized organizational units. Instead of support for small units, private households, and farmers, bigger installations like off-shore wind and solar parks were supposed to become the new beneficiaries (Unnerstall 2017).[5]

---

[5] We will not discuss the issue that the attempts to dampen the expansion of RE were largely unsuccessful in this phase. In 2009 RE accounted for about 16% of the German electricity

This is a phase of vibrant contention especially because the incumbent energy providers were facing severe troubles. The incumbents still tried to defend the status quo by making only piecemeal institutional adjustments. They changed their legitimation strategies again, saying that they now intended to become an active part in the energy transition by providing the backbone for it. Hence, their assessment of RE finally shifted. They intensified their investments in RE. Yet according to a study by Trend Research (Maron, Klemisch, and Maron 2011), only 6.5% of all the renewable plants in Germany in 2010 were operated by the big four companies and most of this share came from hydroelectric power. Nevertheless, the position of the incumbents had definitely changed. They were now picturing themselves as (reliable) partners for the RE actors, as "project enablers" or "system integrators" for prosumers, communities, municipal utilities, or project developers. The competence that RE actors were lacking and the support for the transition period until RE was fully deployed, the incumbents promised to deliver.

Phase three thus ended with a curious situation. On the one hand, the movement for RE had now finally achieved its aim and the switch to the use of RE formally received the support and legitimation of all major parties. The use of the term "transformation" to describe this phase thus seems to be justified, even if the transformation was not the result of conscious planning and strategy. On the other hand, the movement and the support coalition for RE were disintegrating.

4.4. Phase Four (2009–2013): Transformation and Disruptive Innovation: The Break Down of Traditional Legitimation Strategies

In phase four, the generation of electricity based on RE has become the new normal. Having reached a share of almost 25% in 2013, it was no longer

---

supply, in 2011 this share increased to 20%, and at the end of 2013 it was almost at 25% (Unnerstall 2017).

a niche phenomenon, but an integral part of the field of electricity generation. In addition, a thoroughgoing trend towards decentralization has made decisions on electricity-related issues urgent matters of public debate (Unnerstall 2017). Insofar it is no wonder that the RE movement collapsed. A movement needs an opponent against whom it mobilizes. In our case, it was now official government policy to base the energy supply on renewables, and there seems to be no future for nuclear energy. Besides these technology-oriented decisions, a lot of questions nevertheless still require answers. How will a future market operate? Who will be able to participate legitimately in field-related activities? What will be the fate of the old utilities? What will be the role of small-scale installations and their owners? Insofar as layering was an important mechanism in phase two, and open contention was in phase three, they have been replaced in phase four by integration or the successive development of a new market architecture.

An initial aim of the new coalition between the conservative parties (CDU/CSU) and the Social Democrats at the beginning of this phase (2012) was to better influence RE deployment. The government's intention to make RE growth more manageable had failed for several years. The government regularly had to upgrade projections because the growth of installations was much quicker than had been anticipated. In phase four the system for providing support for RE was changed in order to finally stop the quick expansion of RE and to save the incumbents. A new regulatory initiative was passed in 2014, which for the first time explicitly mandated upper limits for the expansion of RE. This was especially targeted at bioenergy and solar energy. Various new instruments were introduced to achieve this aim, among them most prominently the requirement for an actor to participate at an auction in order to obtain a permit to install new capacities. Now only the winners at the auction were to get permission to build installations and connect them to the grid. This measure was accompanied by various other new regulations, which generally were specifically aimed at making life more difficult for decentralized, less professional organizations. Besides the changing political priorities, in contrast to the old incumbents, the actor constellation in the RE field remained rather fragmented. The different types of actors and organizations involved and their different organizational and technological set-ups make it difficult to formulate common goals and to mobilize for them. The lack of hierarchy in the field reflects the inequality in the initial distribution of resources among RE actors. Civil society actors were moving to other arenas to deal with climate change.

Nevertheless, the one development in this phase that most clearly demonstrates that things had changed has surely been the organizational breakup of the two major utilities EON and RWE. In order to increase their chances for survival in a new environment, both companies have split their activities into two parts, one becoming responsible for the old activities (e.g., power plants), while the other one was supposed to tackle new issues such as RE. This should allow for more differentiated and successful learning processes. While the companies responsible for the old stuff were lobbying for capacity markets and the public financing of so-called reserve power plants (to be paid for by the surcharge on RE), the new organizations were busy expanding cooperation with actors in the RE field. In spite of the fact that the incumbents always favor bigger and centralized solutions, the present setup of RE installations makes it necessary for the incumbents to cooperate with local, urban, and/or regional actors. The competence needed to participate successfully in auctions for solar and wind installations can be better garnered by the big organizations, but in the auctioning process local approval of the respective installations is required, which is hardly possible without engaging in some kind of cooperation with locally based organizations.

As far as legitimation strategies are concerned surviving on the new market and within the new regulatory framework have become hot issues. Ecological concerns as such do not play anymore a significant role. Due to the new regulatory framework, economic considerations have become dominant even for primarily ecologically minded actors.

CONCLUSION

There seems to be a wide spread consensus that an energy transition is required to cope with the catastrophic consequences of climate change. How a real energy transition actually will or has to proceed is difficult to assess. The article started with the assumption that there is no such a thing like a unified energy transition. We can combine various developments under the heading of an energy transition, but the actors participating in such practices might have very diverging ideas and plans and find themselves in very different situations. We therefore suggest to look at real instances of change or resistance to change to get a better understanding of what is at stake in a transition process. We conjecture that real actors try to deal with perceived problems in a way that either tries to defend the status quo or changes it. This is of course a simplification, but we might be able to get a clearer analytical view by making these suggestions. Defenders of the status quo and challengers have diverging positions within a field, which develops around attempts to deal with problems. Both groups try to legitimate their activities either with references to the benefits of the status quo or the negative consequences of established practices for the society at large. Meaning that defenders tend to link their legitimation strategies to field-internal justifications (e.g., security of supply), while challengers tend to link up to field external developments (e.g., climate change). Fields are constantly in flux, new ones are emerging, some mature ones are disappearing. The type of legitimation strategy adopted has to be seen in relation to the field context in which they are formulated. The empirical part of the paper provided a sketch of the German system of electricity generation and distribution. It analyzed how legitimation strategies of incumbents and challengers have developed in line with the overall field development, which featured a delegitimation of the once dominant strategies. The structural and regulatory changes outlined above provided the *opportunity* for RE activities to emerge, but they did not create the *necessity* for any one particular form of organization to persist and dominate. Still today, there is not one dominating organization in the field of electricity generation and distribution.

REFERENCES

Abbott, A. (2001). *Time Matters: On Theory and Method*. Chicago, IL: University of Chicago Press.

———. (2016). *Processual Sociology*. Chicago, IL: University of Chicago Press.

Berlo, K., and Wagner, O. (2013). *Stadtwerke-Neugründungen und Rekommunalisierungen: Energieversorgung in kommunaler Verantwortung*. Sondierungsstudie. Wuppertal: Wuppertal Institut für Klima, Umwelt, Energie GmbH. [*New foundations of urban utilities and re-communalization: energy provision under local responsibility. An exploratory study*. Wuppertal: Wuppertal Institute for Climate, Environment and Energy].

Bontrup, H.-J., and Marquardt, R.-M. (2010). *Kritisches Handbuch der deutschen Elektrizitätswirtschaft: Branchenentwicklung, Unternehmensstrategien, Arbeitsbeziehungen*. [*Critical handbook of the German electricity economy: sectoral development, industrial relations*]. Berlin: Edition Sigma.

Bundesnetzagentur. (2007). *Monitoringbericht 2007*. Bonn: Bundesnetzagentur für Elektrizität, Gas, Telekommunikation, Post und Eisenbahnen. [*Federal network agency. Monitoring report 2007*. Bonn: Federal network agency for electricity, gas, telecommunications, posts and railways].

David, R. J., W. D. Sine and Haveman, A.A. (2013). Seizing Opportunity in Emerging Fields: How Institutional Entrepreneurs Legitimated the Professional Form of Management Consulting. *Organization Science*, 24 (2): 356-377.

Dacin MT, Goodstein, J. and Scott, W.R. (2002). Institutional theory and institutional change. *Acadademy of Management Journal* 45:45-57.

Europäische Kommission. (2008). *Spezial Eurobarometer 300: Einstellungen der europäischen Bürger zum Klimawandel*. [European Commission. *Special edition Eurobarometer 300: Attitudes of European citizens and climate change*].Brüssel: TNS Opinion & Social.

Fligstein, N., and McAdam, D. (2011). Toward a General Theory of Strategic Action Fields. *Sociological Theory* 29 (1): 1–26.

———. 2012. *A Theory of Fields*. New York: Oxford University Press.

Hirschl, B. 2008. *Erneuerbare Energien-Politik: Eine Multi-Level Policy-Analyse mit Fokus auf den deutschen Strommarkt*. [Renewable energies Policies: A multi-level policy analysis focussing on the German electricity market]. Wiesbaden: Springer VS.

Hoppmann, J., Huenteler, J. and Girod, B. (2014). Compulsive Policy-Making: The Evolution of the German Feed-In Tariff System for Solar Photovoltaic Power. *Research Policy* 43 (8): 1422–41.

Jacobsson, S., and Lauber, V. (2006). The Politics and Policy of Energy System Transformation: Explaining the German Diffusion of Renewable Energy Technology. *Energy Policy* 34 (3): 256–76.

Jones G, and Khanna, T. (2006). Bringing history (back) into international business. *Journal of International Business Studies* 37: 453-468.

Kriesi, H. (2004). Political Context and Opportunity. In *The Blackwell Companion to Social Movements*, edited by D. A. Snow, S. A. Soule, and H. Kriesi, 67–90. Malden, MA; Oxford, UK; Carlton, AU: Blackwell Publishing Ltd.

Lauber, V., and Jacobsson, S. (2016). The Politics and Economics of Constructing, Contesting and Restricting Socio-Political Space for Renewables – the German Renewable Energy Act. *Environmental Innovation and Societal Transitions* 18: 147–63.

Lawrence T. B., and Phillips, N. (2004). From Moby Dick to Free Willy: Macro-cultural discourse and institutional entrepreneurship in emerging institutional fields. *Organization* 11:689-711.

Lazonick, W. (2005). The Innovative Firm. In *The Oxford Handbook of Innovation*, edited by J. Fagerberg, D. C. Mowery, and R. R. Nelson, 29–55. Oxford: Oxford University Press.

Maron, H., Klemisch, H. and Maron, B. (2011). *Marktakteure Erneuerbare – Energien - Anlagen in der Stromerzeugung*. [Market actors renewable energies. Installations for the production of electricity]. Köln: KNI - Klaus Novy Institut.

Martin, J. L. (2009). *Social Structures*. Princeton, New Jersey: Princeton University Press.

Mautz, R., Byzio, A. and Rosenbaum, W. (2008). *Auf dem Weg zur Energiewende : die Entwicklung der Stromproduktion aus erneuerbaren Energien in Deutschland; eine Studie aus dem Soziologischen Forschungsinstitut Göttingen (SOFI)*. Göttingen: Univ.-Verl. Göttingen. [*On the way towards an energy transition: the development of electricity production from renewable energies in Germany. A study from the institute for sociological research Göttingen/SOFI*]

McAdam, D. and Schaffer Boudet, H. (2012). *Putting Social Movements in Their Place: Explaining Opposition to Energy Projects in the United States, 2000–2005*. Cambridge, UK: Cambridge University Press.

Meyer J. W. and Rowan, B. (1977). Institutionalized organizations: Formal structure as myth and ceremony. *American Journal of Sociology* 83: 340-363.

Neukirch, M. (2010). Die internationale Pionierphase der Windenergienutzung. [*The international pioneer phase of wind energy utilization*]. Göttingen: Georg-August-Universität Göttingen.

Padgett, J. F., and Powell, W.W. eds. (2012). *The Emergence of Organizations and Markets*. Princeton, NJ: Princeton University Press.

Scheer, D., Wassermann, S. and Scheel, O. (2012). Stromerzeugungstechnologien auf dem gesellschaftlichen Prüfstand: Zur Akzeptanz der CCS-Technologien. [Technologies for electricity production in the public discussion: the acceptance of the CCS-technology] In *Akzeptanzforschung zu CCS in Deutschland: Aktuelle Ergebnisse, Praxisrelevanz, Perspektiven* [*Acceptance research on CCS in Germany: current results, practical relevance, perspectives*], edited by K. Pietzner and D. Schumann. München: Oekom-Verlag.

Suck, A. (2008). *Erneuerbare Energien und Wettbewerb in der Elektrizitätswirtschaft. Staatliche Regulierung im Vergleich zwischen Deutschland und Großbritannien*. [*Renewable energies and competition in the electricity economy. Political regulation in Germany and Great Britain. A comparison.*] Wiesbaden: Springer VS.

TNS Emnid. (2013). *Emnid-Umfrage zur Bürger-Energiewende. Ergebnisse einer repräsentativen Meinungsumfrage des Forschungsinstituts TNS Emnid im Zeitraum 23.09.–25.09.2013 im Auftrag der Initiative "Die Wende – Energie in Bürgerhand."* [Emnid-survey on the citizen energy transition. Results from a representative survey conducted by the research institute TNS Emnid between 23.09 and 25.09 2013 at the request of the initiative "The transition – energy in the hands of citizens"] Berlin: Bündnis Bürgerenergie e.V.

Unnerstall, T. (2017). *The German Energy Transition: Design, Implementation, Cost and Lessons.* Berlin Heidelberg: Springer-Verlag.

Ventresca M. J., and Mohr, J.W. (2002). Archival research methods. In *The Blackwell Companion to Organizations.* 805-828 edited by J. A. C. Baum, Cambridge, MA.: Blackwell

White, H. C. (2008). *Identity and Control. How Social Formations Emerge.* 2nd ed. Princeton, NJ: Princeton University Press.

In: A Comprehensive Guide to Energy ... ISBN: 978-1-53616-728-3
Editor: Shannon Alvarado © 2020 Nova Science Publishers, Inc.

Chapter 4

TECHNO-ECONOMIC ANALYSIS OF STANDALONE HYBRID ENERGY SYSTEM

Shweta Goyal[1,*], *Sachin Mishra*[2,†], *and Anamika Bhatia*[1,‡]

[1]Department of Electrical Engineering Graphic Era University, Dehradun, Uttarakhand, India
[2]School of Electronics and Electrical Engineering, Lovely Professional University, Phagawara, Punjab, India

ABSTRACT

Consumption of power is increasing day by day so the more power generation requirement is necessary. This generation cannot be handling by non-renewable sources individually; because they will deplete after certain period of time. So the some gap between power supply and demand can be cover-up by the renewable sources. Renewable system is very useful

---

[*] Corresponding Author's E-mail: shwetugoyal@gmail.com.
[†] Corresponding Author's E-mail: smishra28281@gmail.com.
[‡] Corresponding Author's E-mail: anamikajain2829@gmail.com.

for off-grid system for remote area to build without having complicated grid system.

Rural electrification through main grid increases investment cost and losses. This problem can be overcome by renewable standalone system. The problem with the renewable sources is that they not able to give continue energy so hybrid energy system can be used to decentralize the problem. The term hybrid system describes by the combination of two or more renewable sources with fossil fuel power diesel/petrol to provide continue energy. Hybrid energy system is very useful in location where extension of electrical grid is not possible and it may say very difficult.

Hybrid energy system has found much wider distribution than just as individual stand-alone renewable system for electrification of rural area. Hybrid system is very useful for remote area. But the planning of rural electrification is very difficult; it needs very deep economic, technical and social study of the location. It is very important to develop a proper simulation model and optimization techniques. Wide range of different configuration is possible but choice of modeled configuration must suit the location. It is necessary to have the knowledge about energy demand and resources available for specific location MATLAB script is developed which calculates the optimum system design and allows the user to evaluate the electro-economic and technical feasibility of a large number of technologies. The basic principle is to minimize the total cost of energy (COE) of the system while satisfying the unit and system constraints such as number of solar modules, diameter of parabolic reflectors, number of solar collectors, gross area available hub height of wind turbine, rotor diameter of wind blades, scale factor, area of the installation.

There are a huge number of MPPT techniques and algorithms have come in to existence as discussed in the literature. Most commonly employed technique among them is P&O algorithm due to its simplicity and robust nature. Proposed Artificial BEE colony (ABC) inspired MPPT algorithm has therefore been compared to P&O algorithm. For this new Swarm intelligence based Artificial Bee Colony (ABC) algorithm has been used to track MPP of PV module. Objective function is based on voltage power characteristics of PV module. The constraints taken under consideration include power of the PV module and battery. ABC and P&O aim at tracking maximum power output and minimizing the objective function to global optimal result. Algorithm traces global maxima and minima by voltage versus power curve. Power balance constraints for any period includes power supplied by the hybrid generation and load demand, i.e., generated power must meet total load demand, satisfying power quality and reliability constraints.

Comparative curve for MPPT techniques using ABC and P&O using power curve for PV irradiance shows that ABC algorithm has better power yield as compared to P&O. P&O is the most robust and simple in application but lacks exploitation of PV power at higher irradiance. At normal irradiance below 650 W/m², P&O has been found to exhibit better results while for higher irradiance, ABC gives higher power output.

Generation unit and storage unit capacity depends on load demand. Load data is the actual load as obtained from energy meter reading for the months of the year 2014. Hourly load pattern has been observed for typical days and hourly load profile has been fabricated accordingly. Daily base load was estimated to be 132 kWh/day and peal load of around 35 kW with day-to-day random variability of 15, time step random variability of 20% and load factor of 0.3. Annual average load found to be 132 and peak load as 28.9 kW in the month of August. The obtained hourly load profile during months of year 2014.

No fossil fuel based power source has been considered in this study and thus renewable fraction is .07% but when no fossil fuel used renewable fraction is 100%.The COE found in optimization result is Rs.45.46 and Rs 19.45 in case study 1 (with fossil fuel) and case study 2 (with fossil fuel) respectively, which is less than present COE (grid connection and diesel generator) of Rs. 20.20 and likely to increase in near future as per decreasing COE trend of RES based generation. Obtained COE is competitive COE with hybrid PV-wind-bio-diesel system Rs. 19.45/kWh, hybrid PV-Wind-Diesel Rs. 13.46/kWh, hybrid PV-bio-diesel Rs. 17.71/kWh, hybrid wind- diesel Rs. 14.21/kWh and hybrid wind –bio-diesel Rs. 33.23 /kWh.

Keywords: software, designing, artificial bee colony, conclusion

1. INTRODUCTION

Energy is the key for development of any country as it plays a vital role in sectors like agriculture, industry, transport, commercial and domestic. There are two types of sources of energy generations: Non-renewable and renewable. The sources which cannot be rehabilitated or regenerated rapidly are called nonrenewable sources; while energy which can be repeatedly generated called renewable sources. Many countries are dependent on non-renewable energy because of extra consumption of electricity; but these sources are limited, expensive and will be exhausted after a certain period of time. The maximum use of non-renewable sources may increase several environmental problems on global scale. Now a day's everyone focus on renewable sources because today world is facing massive environmental change due to the different climatic condition. One of the major causes of renewable sources is depletion of fossil fuel which have made focus in the area of energy generation. Some benefits of Renewable Energy (RE) are:

- Its generation improves public health and environment quality.
- It reduces climate changes.
- It also reduces the dependency on fossil fuels gas and oil reserves.
- It facilitates electrification of rural areas

The high cost of renewable energy generation is its main barrier of this field and it can be taken care by couple of optimization techniques which are in use these days. Electricity used for all major domestic and industrial purposes; raises the living standards and increase work efficiency. Electricity can prove to be the best invention of mankind if it is used smartly. It is quite impossible to work without electricity.

At present rural electrification is still a far cry due to the lack of transportation. India's energy crisis may be attributed to the fact that a major share of its rural population is poor; and has less knowledge of updated energy services. Energy has become necessary to do the work of daily life. So Renewable energy plays a vital role to fulfill the energy requirement of any nation.

1.1. Rural Electrification and Its Need

1.1.1. Rural Electrification

To ensure continuous and efficient power in remote areas for domestic and industrial work such as threshing, milking, and hoist grain for storage is called Rural Electrification. Due to the shortage of labor this will increase the productivity at low cost. [Ministry of Power vide letter No. 42/1/2001-D (RE) dated 05.02.2004 have notified the revised definition of village electrification as under]: "A village would be declared as electrified if:

- Basic infrastructure such as Distribution of Transformer and lines are provided in the inhabited locality as well as Dalit Basti/ hamlet where it exists.
- Electricity is provided to public places like Schools, Panchayat Office, Health Centers, Dispensaries, Community centers etc.
- The numbers of households electrified are at least 10% of the total number of households in the village."

Table 1 represents19 state wise status of rural electrification as on 23/12/16[1]. More than 70% population lives in rural areas in India where the connection of grid is not possible therefore some stand-alone system for electricity generation is needed to provide the electricity for unelectrified areas.

1.1.2. Needs of Rural Electrification

Increase the domestic economic levels through the delivery of energy services to fulfill the needs of cookery and illumination.

- Increase venture competence.
- Reduce labor and time in fetching fuel-wood and water.

Energy is required for many basic needs in every sector like irrigation and fertilization, household lightning, food processing, cooking, in small industry processing, in commercial area i.e., shop, flour mills, social services, water pumping, road lights etc.

I. Agriculture Sector- Irrigation, Fertilization.
II. Domestic Sectors: Lightning, Food Processing, Cooking.
III. Industry Sector: Machinery, Mills, Commercials Space.
IV. Social Service: Water Pumping, Health Centre.

---

[1] Press Information Bureau Government of India Ministry of Power.

Table 1. State-wise data on rural electrification, as on 23.12.16

	States/Uts	Total Un electrified	Electrified Village
1)	Arunachal Pradesh	1578	348
2)	Assam	2892	1808
3)	Bihar	2747	2111
4)	Chhattisgarh	1080	553
5)	Himachal Pradesh	35	28
6)	Jammu & Kashmir	134	32
7)	Jharkhand	2525	1397
8)	Karnataka	39	7
9)	Madhya Pradesh	472	358
10)	Manipur	276	185
11)	Meghalaya	912	670
12)	Mizoram	58	30
13)	Nagaland	82	28
14)	Odisha	3474	1908
15)	Rajasthan	495	400
16)	Tripura	26	15
17)	Uttar Pradesh	1529	1459
18)	Uttrakhand	76	7
19)	West Bengal	22	10
	Total state	18452	11363

1.2. Electrification in Indian Context

The data indicates the empowerment in power supply during the year 2016-17. The gap between demand and supply of energy has been reduced to 0.7% from 2.2%. the consumption of electricity is increasing at a fast rate in India as compared to other countries of the world due to increase of population to a larger extent and economic development of our country.

The Figure 1 significantly shows that there is maximum consumption of energy in the eight years as compared to its production. It shows that the consumption of energy is increasing rapidly as compared to its production.

Figure 2 presents the current status of Installed Capacity of India for total energy generation production with Non-renewable and renewable energy sources. It shows that the major portion of electricity generation depends on the non-renewable sources (thermal etc.) and only 18% of total generation depends on renewable sources. The Installed capacity of the country as on dated 30/September/2017 is estimated to be 3,31,117.58 MW[2] comprising 6,780 MW (2%) Nuclear, 44,765 MW (14%) hydro. The capacity of renewable energy in India is rising progressively. In March 2016, RE installed capacity stood at 60,158 MW which was 18% of the power mix in the country (Central Statistics Office, 2017).

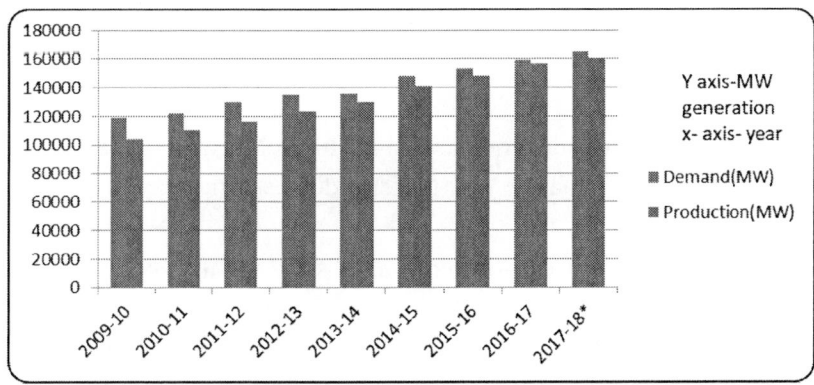

Figure 1. Energy production and consumption per capita in India.

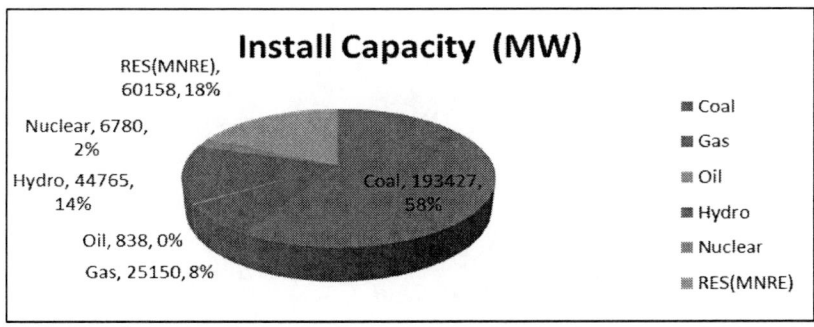

Figure 2. Current installed capacity.

Figure 3 shows that generation through the renewable energy comprise of 29151.29 MW from Wind energy 4346.85 MW from small Hydro Plants, 8182 MW from Biomass Power and Biomass Gasifies, and 9566.66 MW

---

[2] Source: Central Electricity Authority (CEA) Accessed On (18/January/2018).

from Solar power as well as Urban and Industrial waste. India ranks fourth in the world in terms of installed capacity of wind turbine power plants[3]. It shows that the large amount of power generation depends on wind energy system in renewable source.

India is one of the leading developing countries in the world and many investors are interested to invest money. Ministry of New and Renewable energy (MNRE) investigates the matter related to energy generation. MNRE has developed new renewable energy policies to fulfill the energy demand in India. The mission of MNRE is:

1) To decrease the dependency on fossil fuel through development of renewable alternative and thereby decrease the gap between load and demand.
2) To increase the use of renewable sources like bio, wind, solar, hydro, tidal etc.
3) To increase the energy availability in rural areas especially for domestic purpose like cooking, heating etc.
4) To provide affordable or reasonable energy
5) To provide energy equity

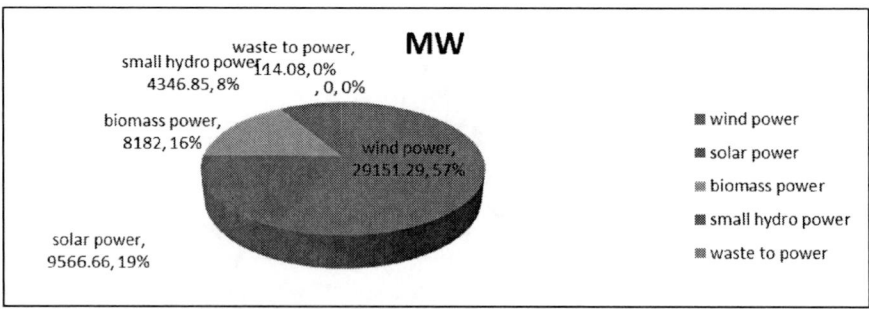

Figure 3. Renewable energy generations in India.

---

[3] Ministry of new and renewable energy (MNRE) (2/January 2018).

Basically, the central and state governments play a very significant part in promotion and development of renewable energy. Planning commission which heads the group of five ministers; works on energy policy sector; state governments have the power to create energy policies directly or indirectly.

1.3. Power Sector Status in India

India capacity of renewable energy is increasing rapidly. The total Installed Capacity of the country as on 30/September/2017 estimated to be 3,31,117 MW[4]. India capacity of renewable energy is rising progressively, and the below Table 2 represent the region wise installed capacity in India.

Table 2. All India installed capacity of power stations

Regions	Hydro (MW)	Thermal				Nuclear (MW)	RES (MNRE)	Total (MW)
		Coal (MW)	Gas (MW)	Diesel (MW)	Total (MW)			
NR	19423.7	52489.2	5781.2	0.0	58270.4	1620.0	12279.1	91593.3
WR	7447.5	69908.6	11059.4	0.0	80968.0	1840.0	18825.3	109080.3
SR	11748.0	43382.0	6473.6	761.5	50617.2	3320.0	27728.3	93413.6
ER	4834.1	27126.6	100.0	0.00	27226.6	0.0	1027.3	33088.0
NER	1312.0	520.02	1336.0	36.0	2292.0	0.0	285.4	3889.4
Island	0.00	0.00	0.00	40.05	40.0	0.0	12.10	52.1
Total	44765.4	193426.5	25150.3	837.63	219414.5	6780.0	60157.6	331117.5

1.4. Rural Electrification Options

There are many types of different ways for rural electrification and some are listed below:

---

[4] Source: Central Electricity Authority (CEA) Accessed On (18/January/2018).

I) *Grid Power Extension*

Grid extension can be the best option for rural electrification as well as village electrification. But it required a huge infrastructure; also, it is very expansive due to following reason.

- Number of users in per kilometer is very small, so consumption is also very less than urban area.
- Ability to pay the electricity bills is very low.

II) *Power Generation with Diesel Generation*

This is the most common conventional technology used in village area. There are many flaws which can be observed and mention below

Some villages are unable to get proper transportation as transportation of diesel becomes very difficult. Which generally increase the transportation cost

- Storing fuel is very difficult.
- High maintenance cost
- Diesel generator increases the effect of greenhouse gases rapidly.
- Rise in noise pollution level.

III) *Nonrenewable Energy System*

The sources which cannot be rehabilitated or regenerated rapidly are called nonrenewable sources. Many developed nations need nonrenewable energy sources such as fossil fuels (coal and oil) and nuclear power. More than 85% of the energy is used in the form of nonrenewable supplies. Problem for Rural Electrification through Nonrenewable Sources listed below:

- Remote or rural communities must import the primary fuel for power production, which increases the transportation cost
- Cost of storing the fuel is very high
- Some Non-conventional resources produce greenhouse gases.

- These systems are noisy

IV) *Renewable Energy System*

Renewable can be defined as energy sources for the long term for daily life works. It depends on earth's ecosystem like geothermal, solar system etc. Today India is developing the production of power from sources like solar, wind, biomass and hydro. Renewable energy sources are commonly used to reduce the gap between production and demand all over the world.

Characteristics of renewable energy are as follows:

a) Less noisy.
b) Easily accessible.
c) Less expensive.
d) Less maintenance.
e) No waste product and no pollution.
f) No transportation system required.
g) Difficult to generate the high power
h) Difficult to get continuous energy because of dependency on natural sources.
i) Requires battery storage, converters etc.

One possible solution to overcome the drawbacks of renewable and non-renewable technologies to employ both types of combination to minimizing the cost of the system. As a result, energy generation from individual renewable sources cannot be reliable or continuous. So other possible solutions that may overcome the drawbacks of an individual renewable energy system are some stand-alone system-

a) Hybrid energy system
b) Integrated energy system

Flow chart of rural electrification methods is shown in Figure 4.

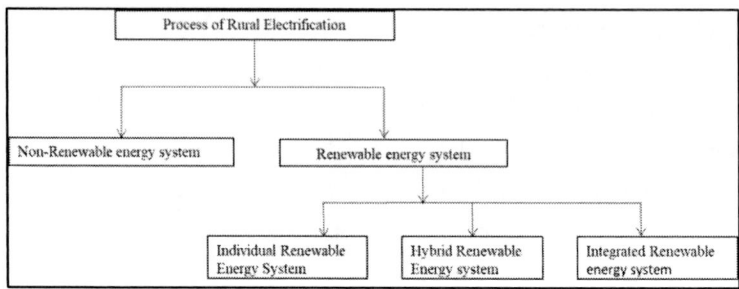

Figure 4. Rural electrification option.

1.5. Component for Hybrid Energy System

Primary Sources- renewable resources like solar, wind, biogas, biomass, hydroelectric, geothermal hydrogen fuel cell, etc. useas a primary source.

Secondary Resources – non-renewable energy resources like coal, petroleum and natural gas are fossil fuels which took millions of years to be formed and cannot be renewed during our life span.

Battery System – Battery System is useful to store the excessive charge for later use. If primary and secondary sources are not available than we can use that battery system for energy

Control Unit – Control unit includes inverter (for dc to ac converter), ac to dc converter for storage system, amplifier to increase the amplitude, filter etc.

Dump Load – A dump load is provided to save the battery from excessive charging .it is used for safety purpose.

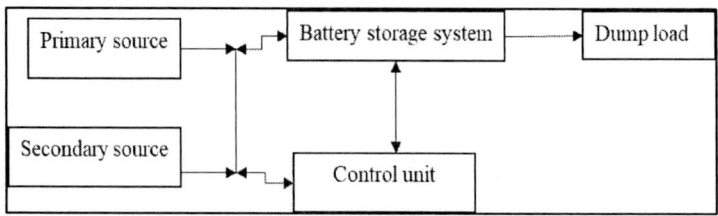

Figure 5. Components for hybrid energy system.

1.6. Different Configuration of HRES

Different configurations of hybrid energy system based upon the different connection between the components are as follows

I) *Series Hybrid Energy System* – A series hybrid energy system shown in figure, either conventional source or non-conventional source is used to charge the battery bank. Diesel generator connected to the battery charger. Inverter is used to convert DC power to AC power. Characteristics of series HRES are follows:

- System efficiency is low
- Large size of battery and inverter

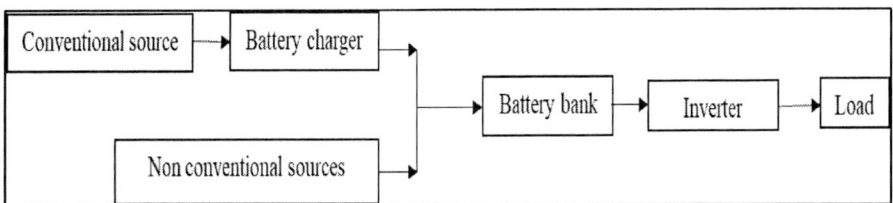

Figure 6. Series hybrid energy system.

II) *Alternate Hybrid Energy System* – In alternate HRES; conventional sources are used at the time of peak load or when any renewable sources are not available at that time load can be supplied directly from the conventional source. Non-conventional sources can charge the battery for low demand and during the low demand of electricity conventional sources are switched off and the load supply through the battery shown in Figure 7. Characteristics of Alternate HRES are follows:

- Can also operate at night.
- Improved system efficiency.

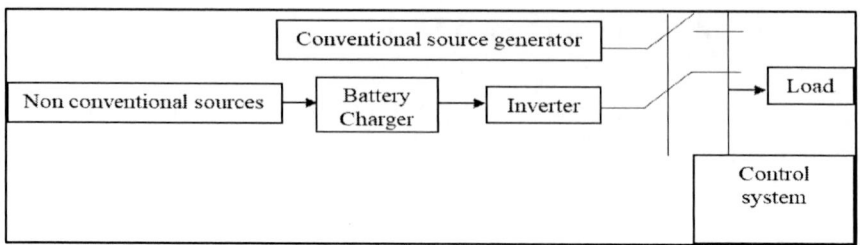

Figure 7. Alternate hybrid energy systems.

III) *Parallel Hybrid Energy System* – In parallel hybrid energy system diesel generator and inverter run parallel shown in Figure 8.

Characteristics of parallel HRES are follows:

- Highly optimized
- Most efficient
- Requires minimum maintenance

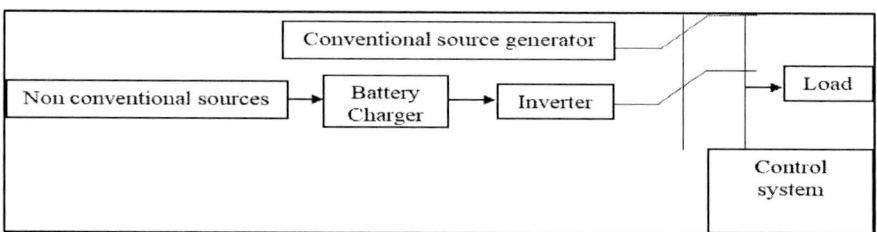

Figure 8. Parallel hybrid energy systems.

2. TOOLS SELECTION FOR FEASILBLE ANALYSIS

There are so many academic accessible software tools which can be used to find out the technical-& Economical possibility of different energy system as photovoltaic, wind, and biomass. Based on some literature reviews there are so many potential software tools to simulate and provide optimize output like HOMER, EnergyPLAN, TRNSYS and Hybrid2. Further reviews of different kind of tools are described in Table 2.5; accounting their advantages, disadvantages and available feasibility options; HOMER software has been found to be the most suitable for this study. This software performs on technical part; rather than carrying out time-series simulations like optimization of cost optimization and sensitivity analysis. TRNSYS, Energy PLAN and Hybrid2 can optimization operation but not woke on cost and sensitivity part. Additionally, the HOMER has validated natural algorithms which used to synthesize irradiance and wind speed data if experiential data is not obtainable. In new possibility outlook, this software has appeared to be a manufacturing standard rules and regulations for modeling a HRES.

2.1. Homer Software Tools

Hybrid Optimization Model for Electric Renewable (HOMER) is copyrighted of Midwest Research Institute (MRI) and its computer model has been developed in US National Renewable Energy Laboratory (NREL). It uses to design and compare the power systems generations with other worldwide technologies. Performance can be simulated by comprise a mixture of Non-renewable generators, collective temperature and energy, bio generators, wind turbines, photovoltaic cells, numbers of battery, hydro turbine, flywheel, fuel cells, power converters. It has three maintasks, like simulation, optimization and sensitivity analyze. User gives the input. The performance of an individual power system pattern for per minute (5,25,600 lines) to per hour (8760 lines) of the year is modeled to determine its technical feasibility and lifecycle cost. In the simulation process two purposes are served; firstly, to determine the feasibility in the life cycle cost, second the total install and operating cost.

HOMER legacy interface for set up the power system model is shown in Figure 9. Selected system can be used to select the different component to explain individual category of loads for a preferred configuration of power system. Input data for prescribed system configuration are primary energy resources data, cost data, interest rate on capital etc.

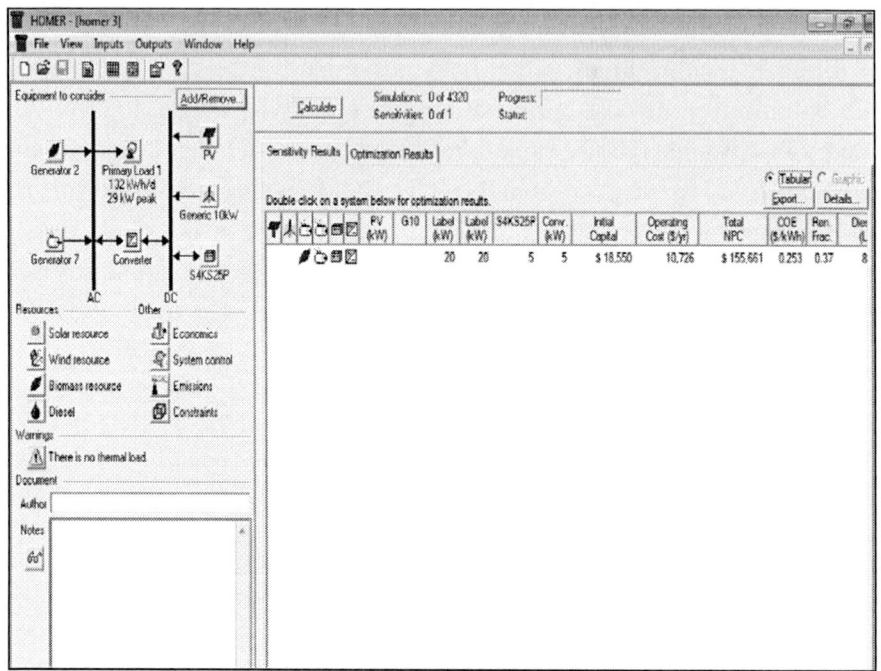

Figure 9. Homer system component selections.

2.2. Introduction of MATLAB

To manage a reliable and economical system for long time; it is very essential to choose the best combination of component and dimensions of the hybrid energy system. More financial investment needed for large number of resources, while a system with a small number of components may result in the interruption of electricity supply in the electricity system.

Climatic conditions also affect the choice of selection of renewable energy sources. For example, PV hybrid systems are best suitable in areas with warm climates and in areas where there is large number of sunny hours. Proposed system included PV generator, wind turbine, batteries, biomass generator with diesel generator and control variables.

Step by step designing of a Photovoltaic system/Wind turbine/biomass hybrid renewable energy system (HRES) with battery and convertor system but without any grid connection with the help of MATLAB/SIMULINK software has been discussed. MPPT (Maximum power point tracker) system for PV (Photo voltaic system) generated by Perturb and Observe algorithm to maximize the power generation. Active performance of the system observes in dissimilar environmental circumstances for one 30 kW load. The solar PV system is designed to generate approx. 45-50kW, the wind system is basically designed for 147kW approx. and biodiesel is also designed for approx. 60kW. The efficiency of the wind system is maximum 40-45% at the peak wind speed practically so it is considered for higher rating. The model is basically designed in the discrete mode with the sample time 20 μs, to maintain a unidirectional transmission a diode is connected across each source and an ideal switch which maintains the switching of the respective sources.

MATLAB script is developed which calculates the optimum system design parameters and allows the user to estimate the electro-economic and technical expediency of a large number of advanced technologies. The main objective is to minimize the total Net Present Cost (NPC) of the HRES system while satisfying the individual system constraints such as number of solar modules, diameter of parabolic reflectors, number of solar collectors, gross area available hub height of wind turbine, rotor diameter of wind blades, scale factor, area of the installation. In determining the appropriate system components ratings following points must be considered:

- The system must reduce the overall cost of production of electricity (Rs / kWh)
- It must be ensured that reliable needs of energy consumers are totally met.

It is very important to optimize the simulated system for best result. Size is a very important part of designing a system. Meaning of size optimization is to optimize the different components of a system to reduce overall the cost.

The aim of a HRES system should be to reduce the use of diesel fuel as much as possible to save the maintenance cost also. Optimization of HRES reduces net present cost with changing the desired output.

Figure 10 shows the diagram of HRES. A basic understanding of three different sources and a battery as an emergency can be employed in an isolated area/village where transmission of electricity is quite costly, and it can be overcome by usage of these two or three sources. From the last few decades scientists are working towards biodiesel which is made from vegetable oils is also replaces the available diesel.

Solar irradiation and solar temperature are used as a input parameter of solar system. This is followed by the MPPT CONTROLLER to track the maximum point of solar radiation. Produced fluctuating DC given to the DC to DC converter to get an ideal DC voltage. Than given to the DC to AC converter for AC load

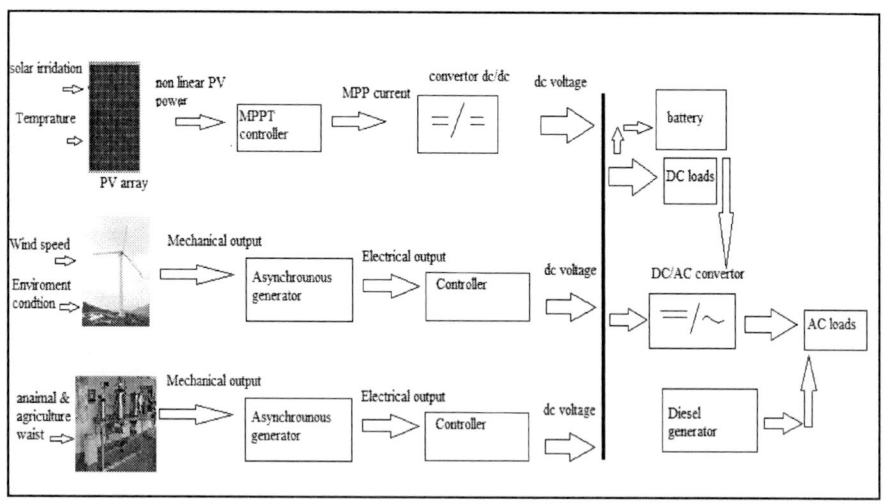

Figure 10. Block diagram of proposed HRES system.

Speed of wind turbine depends upon the wind speed and different environment condition. Some mechanical out is generated through the wind turbine. This mechanical output fed to the asynchronous generator to convert mechanical output into electrical output. Controller sends the dc voltage to DC to AC converter to convert into the signal according to the load.

Animal and agricultural is used for biomass plant. This mechanical output fed to the asynchronous generator to convert mechanical output into electrical output. Controller sends the dc voltage to DC to AC converter to convert into the signal according to the load.

Diesel generator is used as a secondary source, when all these renewable sources are not available than diesel system is used for energy generation to full fill the demand.

3. COMPONENT PARAMETERS

In this work, solar, wind and biomass provide primary power source for load, and battery as a secondary unit. Every individual system is replicated in terms of cost and size to get minimum cost power explanation.

3.1. Solar System

Photovoltaic system consists of a series-parallel module to generate electrical energy. Solar irradiation use as an input parameter. Output calculated from the following equation:

$$P_{PV} = Y_{PV} f_{PV} \left(\frac{G_T}{G_{T,STC}}\right) [1 + \alpha_p (T_c - T_{c,STC})] \qquad (1)$$

Where:

Y_{PV} = Power output [Kw]
f_{PV} = Derating factor [%]
G_T = Solar radiation incident on the PV array [kW/m²]
G_T, STC = Incident radiation at standard test conditions [1 kW/m²]
α_p = The temperature coefficient of power [%/°C]
T_c = PV cell temprature
$T_{c,STC}$ = PV cell temprature standard test condition

IF temperature does not affect the model than Above equation can be simplified as:

$$P_{PV} = Y_{PV} f_{PV} \left(\frac{G_T}{G_{T,STC}}\right) \qquad (2)$$

Monthly clear index representative the fraction of the solar radiation outstanding the peak ambiance that makes it through the environment to strike the ground. The following equation defines the monthly average clearness index

$$K_t = \frac{H_{ave}}{H_{o,ave}} \qquad (3)$$

where:

H_{ave}= Monthly average radiation on the horizontal surface of the earth [kWh/m²/day]

$H_{o,ave}$ = Radiation on a horizontal surface at the top of the earth's atmosphere [kWh/m²/day]

3.2. Wind Generation System

Output power can be calculating by software in three steps; first Calculate the wind speed at the height of hub. Secondly generated power in wind turbine at that wind speed with standard air density. Finally, HOMER adjust the output power for the real air density by apply the power law, the wind speed at hub height can be calculate using the following equation

$$U_{hub} = U_{anem} \left(\frac{zhub}{zanem}\right) \alpha \qquad (4)$$

Techno-Economic Analysis of Standalone Hybrid Energy System

U_{hub} = Wind speed at the hub height of the wind turbine [m/s]
U_{anem} = Wind speed at anemometer height [m/s]
z_{hub} = Hub height of the wind turbine [m]
z_{anem} = Anemometer height
α = Exponentitial law

To regulate to real circumstances, value of power predict by the power curve by the air density ratio, according to following equation:

$$P_{WTG} = \left(\frac{\rho}{\rho_0}\right) P_{WTG,STP} \tag{5}$$

P_{WTG} = the wind turbine power output [kW]
$P_{WTG,STP}$ == the wind turbine power output at standard temperature and pressure [kW]
ρ = the actual air density [kg/m³]
ρ_0 = the air density at standard temperature and pressure (1.225 kg/m³)

3.3. Generator Plant

A generator operates either on fossil fuel and biogas. In every time interval, it calculates the necessary output of the generator and the equivalent mass flow rates of fossil fuel and biogas. This calculation is based on several key assumption:

- Invariable ratio of biogas substitution (zgas), self-regulating output power of engine.
- The systems always use to increases the use of biogas as compare to the use of fossil fuel.
- The fraction of fossil cannot go under the definite smallest amount.

Even if the derating factor is less than 100% connected with operating in dual-fuel mode, the generator can produce up to 100% of its rated power, provided the fossil fraction is maximum.

4. COST COMPONENTS

It includes initial capital cost (C_{icap}), replacement cost (C_{rep}), annualized cost (C_{annu}), salvage value (C_{sal}), net system cost (C_{net}), and unit energy generation cost (C_{unit}).

- Initial capital includes the cost of purchasing the equipment and establishment cost of the system.
- Replacement cost includes replacing a component at the end of its useful life cycle. It may be same as installation cost or differ from initial cost
- O&M cost is sum of major and miscellaneous expenses made for smooth functioning of the system during project tenure.
- Salvage value includes cost of system or component after completing its life. It depends on replacement cost, remaining life of component (Y_{rem}) and rated life of component (Y_{life}).
- Life cycle cost (*LCC*) includes total cost of installation and operation for specified time span. It is important for financial evaluation of hybrid system and given as:

$$LCC = C_{icap} + C_{rep} + O\&M\ cost + fuel\ cost - C_{sal}$$

Other economic consideration includes annualized cost, operating cost, net present cost and levelized cost of energy.

A) *Annualized Capital Cost*
Annualized cost (C_{annu}) can be obtained by multiplying net present cost (C_{NPC}) with function of capital recovery factor ($C_{f,rec}$), given as:

$$C_{annu} = C_{NPC} \times C_{f,rec}\left(i_r, Y_{proj}\right) \qquad (6)$$

$$C_{f,rec}(i_r, Y_{proj}) = \frac{i_r(1+i_r)^{Y_{proj}}}{(1+i_r)^{Y_{proj}} - 1} \quad (7)$$

$$i_r = \frac{i' - f}{1 + f} \quad (8)$$

Where, i_r is annual interest rate, Y_{proj} is project lifetime, i'_r is nominal interest rate (the rate at which loan can be granted) and f is inflation rate.

An optimum combination of a hybrid solar-wind-bio generator system can make the best compromise between the two considered objectives: the system power reliability and system cost. Annualized cost of system is an important part of economic analysis.

B) *Annualized Replacement Cost*

In the studied hybrid system, only the battery needs to be replaced periodically during the project lifetime.

$$C_{annu,rep} = C_{rep} \cdot SFF(i_r, Y_{life}) \quad (9)$$

Where, C_{rep} is the replacement cost of the component (battery), US$; Y_{life} is the component (battery) lifetime, Year; SFF is the sinking fund factor, a ratio to calculate the future value of a series of equal annual cash flows. The equation for the sinking fund factor is:

$$SFF(i_r, Y_{life}) = \frac{i_r}{(1+i_r)^{Y_{life}} - 1} \quad (10)$$

C) *Maintenance Cost*

The system maintenance cost, which has taken the inflation rate f into account, is given as:

$$C_{annu,main}(n) = C_{annu,main}(1) \cdot (1+f)^n \quad (11)$$

Where, $C_{annu,main}(n)$ is the maintenance cost of the nth year.

D) Net Present Cost (NPC)

It is differentiation among current price of each cost over its life span and current price of all returns that over its lifetime. Here costs contains initial capital cost, component replacement cost within project tenure, O&M cost, fuel costs, purchase of power from grid, taxes and penalties. Revenues include salvage value and sales revenue. Thus economic analysis is done by ranking system as per their NPC and assumes flat price rise over project tenure. NPC is given as:

$$NPC = \frac{C_{tot,annu}}{CRF(i_r, Y_{proj})} \qquad (12)$$

Where, $C_{tot,annu}$ is total annualized cost and CRF is capital recovery factor as a function of annual interest rate i_r and project lifetime. CRF is given as:

$$CRF(i_r, Y_{proj}) = \frac{i_r(1+i_r)^{Y_{proj}}}{(1+i_r)^{Y_{proj}} - 1} \qquad (13)$$

E) Levelized Cost of Energy (COE)

It is defined as the average cost per KWh of useful electrical energy generated by system. To calculate it, annualized cost of generating electricity is divided by total useful electrical load served (E_{served}) as given in following equation:

$$COE = \frac{C_{tot,annu}}{E_{served}} \qquad (14)$$

$$E_{served} = E_{prim} + E_{deff} + E_{grid} \qquad (15)$$

Where, E_{prim}, E_{deff} and E_{grid} are toal amount of primary, defferable and grid energy (if any) sold to grid respectively.

F) Salvage Value

Salvage value refers to the remaining value of the component at the end of project tenure. Linear depreciation of components has been assumed i.e., salvage value directly proportional to the remaining life of component. It is also assumed that salvage value depends on replacement cost and not initial capital cost. It is given as:

$$C_{sal} = C_{rep} \frac{Y_{rem}}{Y_{comp}} \tag{16}$$

Where, Y_{rem} is the remaining life of component at the end of project tenure and given as:

$$Y_{rem} = Y_{comp} - \left(Y_{proj} - Y_{rep}\right) \tag{17}$$

Y_{rep}, the replacement cost duration and calculated as:

$$Y_{rep} = Y_{comp} \left[INT\left(\frac{Y_{proj}}{Y_{comp}}\right) \right] \tag{18}$$

Where, INT refers to a function to return integer value of real number.

G) Renewable Fraction (f_{ren})

It is defined as the fraction of electricity fed to load by RES. It is calculated as:

$$f_{ren} = 1 - \frac{E_{nonren}}{E_{ser}} \tag{19}$$

Where, E_{ser} is total electrical energy served and E_{ninren} is electrical energy generated from non-renewable sources.

5. Modeling Method

A systematic modeling method is necessary for the development of an optimized hybrid renewable energy system for remote areas; as it ensures continuous power supply to remote rural households. Important steps of the modeling method for the present study are discussed:

5.1. Site Selection

Dehradun block has adequate sunshine as a natural resource; it has moderate wind speeds while biomass is available in great quantity as the animal population and agriculture area are much greater in this area and nearby.

Dudhali village is selected for this study, located at 30.232110N, 78.014992S in Uttrakhand state. Table 3 gives the details of village. The adjacent town is Mussoorie, which is about 16-19 km away. Dudhaliis located nearly cloud's ENDarea. The village has hilly area and does not have transportation facility. Some houses light lantern for energy.

Table 3. Details of the remote area

Data	Particulars
Name of village	Dudhali
Region	Dehradun
State	Uttrakhand
Nation	India
Latitude	30.232110N
Longitude	78.014992S
Grid electricity	Yes(only 5-6 hour)
Total number of house holds	15
Education facilities	yes
Medical facilities	yes
Post office	no

5.2. Load Assessment

Data collected by the survey and the demand of energy is separated into five main category (i) Domestic Load (ii) Buisness Load (iii) Agriculture/Irrigation Load (iv) School Load (v) Medical Centers. Domestic load consists of light, television set, fans, and radio system. Commercial load includes power for shops, community centre, and street. Agriculture load includes water pump, irrigation pump, well, fodder machine. Medical centre load consists of compact fluorescent light (CFL), refrigerator, fans, while school load consist of CFL, ceiling fan, desktop, television.

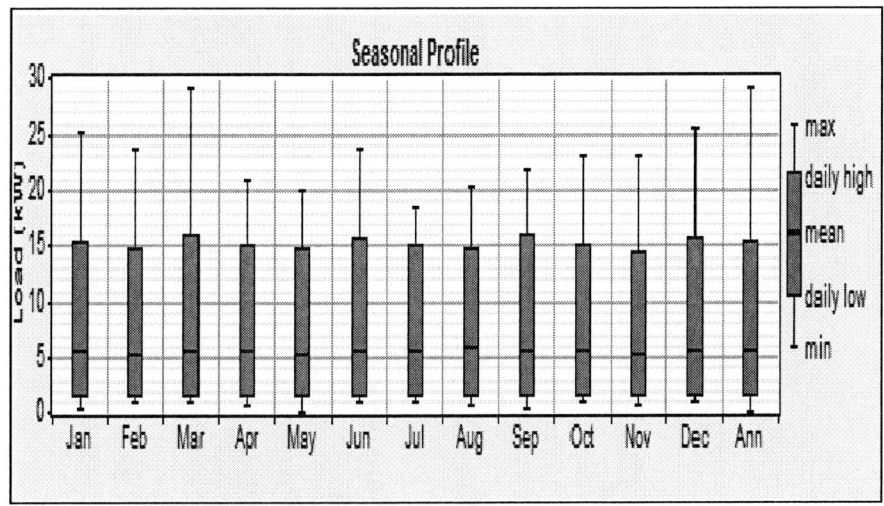

Figure 11. Hourly loads during year 2014.

5.3. Energy Resources

To reduce the cost of the system and to enhance the efficiency of the system, size optimization of different component has been done. The components are solar Photovoltaic system, wind turbine, biomass, diesel generator, converter and battery.

5.3.1. Photo Voltaic System

Table 4. Solar resource input in optimization software

Month	Clear index	Daily radiation (kWh/m2/d)
January	0.620	3.571
February	0.627	4.391
March	0.627	5.429
April	0.637	6.479
May	0.653	7.258
June	0.618	7.075
July	0.519	5.837
August	0.537	5.625
September	0.623	5.698
October	0.707	5.289
November	0.699	4.211
December	0.608	3.256
Average	0.617	5.345

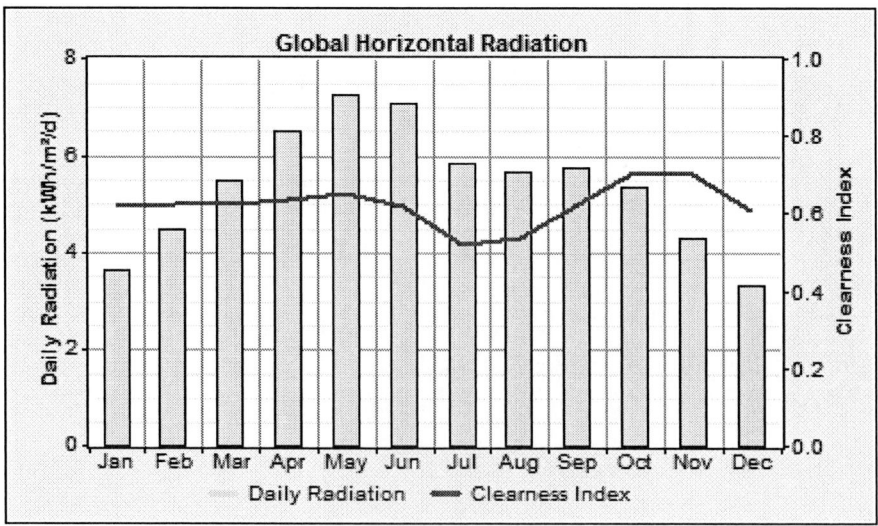

Figure 12. Graphs for solar input.

Photovoltaic module manufactured by 'Luminous' has been used in this study. Each unit has rated power rating of 250 W and consists of 60 cell multi-crystalline solar photovoltaic module. It has module efficiency of

15.60% and use efficiency of 95%.Average estimated irradiance (I_{em}) has been taken as 5.3 KW/m² with ground reflectance of 20%. Derating factor in this study has been considered as 80%. 1KW PV module needs approximately 10-12 square meter of installation area.

The cost for 1 KW system is Rs.51351.38/kw and salvation cost is taken Rs. 15953.16/kw. 10, 20, 50, 100, 200 kW used as the different capacities for Solar PV panels. And system life is considered 20 years.

Table 4 shows the PV input data used in software and Figure 12 shows the graphical representation.

5.3.2. Biomass Energy

The estimation of Biomass probably depends on the accessibility of animal and agriculture waste. The availability of biomass on the site is about 14tones/year, Table 5 shows the biomass data in that area. Some biomass is used for domestic purpose and balance is used for power generation using biomass gasification.

Table 5. Biomass resource input

Month	Available biomass(tones/year)
January	10.000
February	15.000
March	17.000
April	10.000
May	15.000
June	17.000
July	10.000
August	15.000
September	17.000
October	10.000
November	15.000
December	17.000
Average	13.984

5.3.3. Wind Turbine

Wind turbine selection depends upon the wind speed and power curve at that area. shown in table 6. A generic 10 kW WT used at Rs.418890.28cost initially Rs. 104688.34salvation cost with Rs. 359459.63 maintenance cost with 25-meterhub height.

Table 6. Wind turbine resource input

Month	Wind speed(m/s)
January	10.200
February	10.200
March	10.200
April	12.200
May	12.500
June	12.500
July	12.500
August	11.200
September	11.000
October	10.000
November	10.000
December	10.200
Average	11.061

5.3.4. Battery Bank

Selection of size and type of backup unit is among the deciding factors for establishment of standalone unit. Storage unit ensures better reliability and fills energy gap between generation and load demand. Deep cycle battery is commonly used in standalone renewable energy system. Manufacturing specifications of the battery has been given in Table7 Hybrid PV-wind-bio generation reduces the need of backup unit to great extent, reducing overall generation cost.

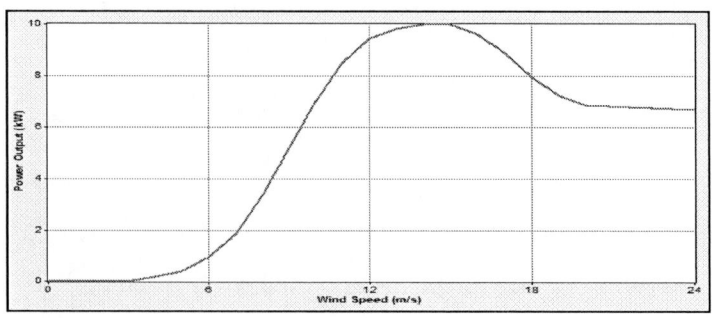

Figure 13. Graphical Representation for Wind Turbine output and wind speed.

Techno-Economic Analysis of Standalone Hybrid Energy System 105

Table 7. Technical data of battery bank

Technical Data	Value
Make	Rolls
Model	Surrette 4KS25P
Voltage (DC)	4v
type	Flooded lead acid
Ampere hours	1347ah
Size and weight	10.5*24.75*15.75/ 315lbs
Life cycle	15

5.3.5. Diesel Engine-Generator Set

A generator of 3 KW AC is used at cost Rs.23142.35 initially and salvation cost Rs. 39437.86 is considered with Rs. 26.02/hour process and preservation charges. Time is predicted as 15000 working hours. Different capacities of ac generators are considered, for e.g.15 kW,20 kW for optimization. Costs of various apparatus are given in Table 8.

This proposed model is prepared with the help of one 10 KW photovoltaic cell, 1 generic 10kW WT one 200 kW bio diesel generator, one 1 KW converter, 1 S4KS25P battery, one 3 kW diesel generator (for backup system) with 533kWh/day and peak load is 90kW shown in Figure 14.

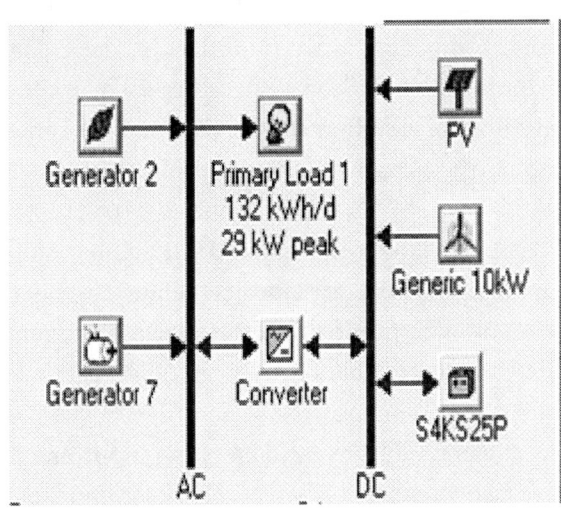

Figure 14. Schematic architecture of proposed hybrid system.

Table 8. Costs of various components

Component	Cost(Rs./kW)
PV cell	58813.62
Biomass generator	65871.26
Diesel generator	7818.75
Wind turbine	311366.25
Convertor	51894.38
Battery	69192.50

5.4. Economic Considerations

The input of economical component includes capital cost, replacement cost, O&M costs, fuel cost, salvage value and lifetime. In this study, considered input economical and component life has been estimated. The life time of the project considered to be 25 years with annual interest rate of 6%. As per manufacturer specification; these components have been replaced in order to complete the project. Useful life time of components has been taken as per manufacturer specification and actual life of component can be more than rated time period.

Search space provides population space to find optimal system configuration and component size in order to meet load requirement economically in terms of NPC. Search space depends on site meteorological condition and installation area limitations. Size and cost optimization result is shown in Figure 15. First row shows results with lowest NPC of Rs 7326951.12 and COE Rs. 11.91, initial capital Rs 1682134.11.

Cash flow has been shown in Figure. 16. Each bar shows expenditure and income from components during project tenure. In this figure, negative bar shows expenditure i.e., outflow of cash. Bar at year zero shows capital cost and consecutive negative bars at 10th and 20th year shows expenditure towards component replacement. At the end of 10th year battery has been replaced while at the end of 20th year battery, wind turbine and converter is replaced. Positive bar shows the salvage value of components at the end of project tenure. Ideal salvage value has been considered and the only

Techno-Economic Analysis of Standalone Hybrid Energy System

components having remaining life as per manufacturer specification is counted.

PV (kW)	G10	Label (kW)	Label (kW)	S4KS25P	Conv. (kW)	Initial Capital	Operating Cost ($/yr)	Total NPC	COE ($/kWh)	Ren. Frac.	Diesel (L)
	1	1	15	5	10	$24,568	6,449	$107,012	0.174	0.69	5,21
1	1	1	15	5	10	$25,318	6,490	$108,280	0.176	0.70	5,05(
1	1		15	5	10	$25,218	7,466	$120,653	0.196	0.66	5,96
1			20	5	10	$26,418	7,890	$127,285	0.207	0.63	6,46(
		20	20	5	5	$18,550	10,726	$155,661	0.253	0.37	8,95
1		20	20	5	5	$19,300	10,922	$158,926	0.258	0.38	8,74;
1			15	5	10	$19,100	14,074	$199,009	0.323	0.03	13,06;
			15	5	10	$18,350	14,189	$199,728	0.324	0.00	13,44;
1	1	20	50		5	$32,118	20,792	$297,907	0.484	0.47	17,55
	1	20	50		5	$31,368	20,867	$298,120	0.484	0.46	17,78
		20	50		1	$22,250	22,388	$308,438	0.501	0.21	19,32
1		20	50		1	$23,000	22,583	$311,688	0.506	0.22	19,32;
50	1			50	50	$131,118	16,481	$341,804	0.555	1.00	
50	1	1		50	50	$131,218	16,520	$342,404	0.556	1.00	
50	1		50		10	$70,618	28,225	$431,432	0.700	0.64	16,33
	1		50		5	$29,368	31,950	$437,797	0.711	0.27	28,52(
			50		1	$20,250	37,346	$497,656	0.808	0.00	33,59(
50			50		10	$64,500	34,043	$499,680	0.811	0.46	21,59(

Figure 15. Size and cost optimization result.

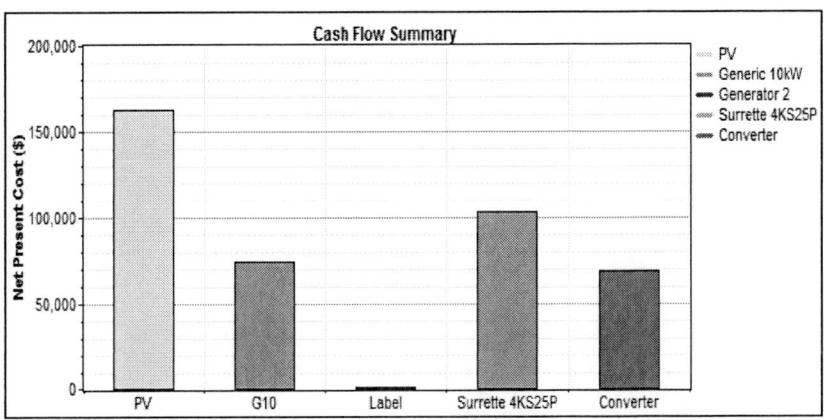

Figure 16. Cost components of equipments involved in project.

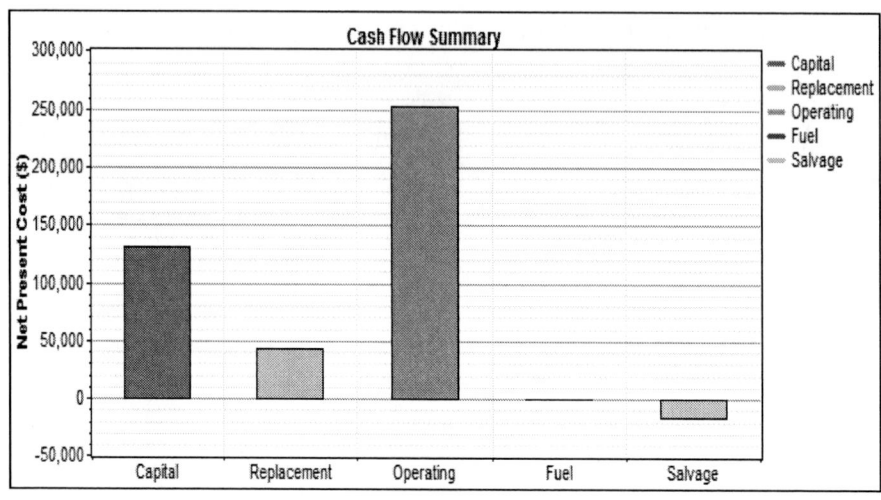

Figure 17. Cash flow due to cost type in project.

Table 9. System cost

Cost	Component	PV ($)	Wind (Generic 10kw)	Generator Bio Mass	Generator or diesel	Battery surrette 4KS25P	Converter	System
Capital Cost ($)	Net	37,500	6,118	100	5,850	50,000	37,500	131,218
	Annualized	2,934	479	8	458	3,911	2,934	10,265
Replacement Cost ($)	Net	11,693	638	455	9,147	29,758	0	42,544
	Annualized	479	50	36	716	2,328	0	3328
O&M Cost ($)	Net	120,164	67113	70	5,122	31,958	31,958	251,263
	Annualized	18	5,250	5	401	2500	2,500	19655
Fuel	Net	0	0	0	51,701	0	0	51,701
	Annualized	0	0	0	4,044	0	0	4,044
Salvage Value ($)	Net	-6553	-119	-2	-380	-8543	971	-16,211
	Annualized	-513	-9	-2	-30	-668	-76	-1268
Total Cost ($)	Net	162803	73,740	600	71,440	103173	16,211	408,814
	Annualized	12,736	5,769	47	5,584	8071	5358	31,980

5.4.1. Electricity Generation

Optimization results give annual production from PV, wind, generator (Bio), Generator (diesel), about 88,275 kW, 48,882 kW and 1,346 kW and 19,473 respectively. Monthly average electricity production has been shown in Figure 18. PV, wind generates and bio generator about 64% and 35% and 1% of total generated electricity.

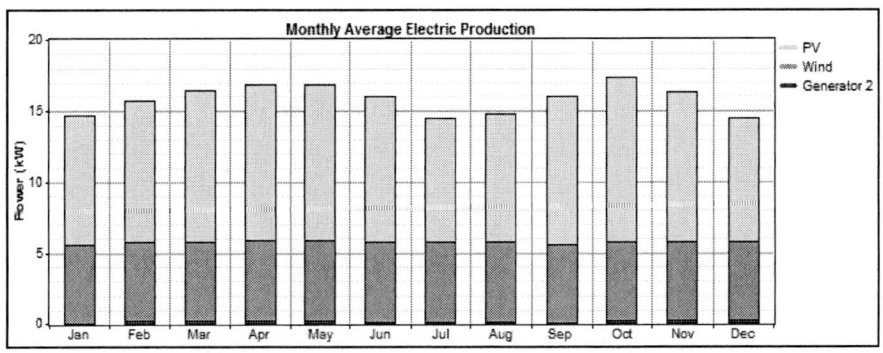

Figure 18. Monthly average electricity production.

6. LITERATURE REVIEW

Zoulias et al. [1] used hydrogen technologies for optimization. Zoulias concluded that the salvation of fossil fuel generator with hydrogen technology was theoretically possible but not economically feasible.

Gupta et al. [2-5] optimized the cost for PV/biomass/biogas/ small/hydro/ battery /fossil fuel HES with the help of dispatched strategy algorithm. To reduce the use of diesel generator for determining the battery storage type HRES for future supply. The authors gave general mythological structure plan for the micro HRES in remote area by proposed 6 stages namely- selecting cluster of villages, demand assessment, Resource assessment, Estimation of unit cost, Sizing and optimization, and model formulation. A numerical example is also included to demonstrate the action plan.

Athima et al. [6] extracted energy from RE and tried to bring the light concept of HRES. Authors prepare a frame of various functions which show the best approach to give power for small grids with the help of tools and technique, integrating RE.

Chin et al. [7] modeled a single axis solar tracker with two light-dependent resistor (LDR) in MATLAB software. It operates at different modes and rotates automatically according to solar irradiance all through sunlight hours, and sleep mode during night time to save the energy. To provide the flexibility one tidy follower used which operate at dissimilar mode to achieve efficiency over the fixed solar panel.

Hafez et al. [8] minimized the life cycle by using diverse case like only diesel, a fully non-conventional based, a diesel-non-conventional mixed, and a grid-connected externally in HOMAR software. Hafez compared their financial, performance and ecological discharge. Author concluded that the diesel-non-conventional mixed micro grid has the minimum Net Present Cost (NPC) and a small carbon footprint when compared to a stand-alone diesel-based microgrid. Although a fully renewable-based micro grid, which has no carbon footprint, is the most preferred, the net present cost (NPC) is higher.

Sen et al. [9] proposed a case study for village PALARI in the state of Chhattisgarh. Residential, institutional, commercial, agricultural load was used for pre-optimization software analysis and best off-grid option was recognized. Comparison was done with grid extension in future.

Fakehi et al. [10] modeled an HRES based on air/electrolyzer /PEM fuel cell for Khaf region-Iran. The model was based on the thermodynamic, electrochemical and mechanical model. Authors concluded that the power effectiveness and energy coefficient is maximum at optimum wind speed.

Chang et al. [11] designed a PV/wind/ diesel/battery HRES in Monte Carlo simulation. Authors used simulation-optimization techniques for HRES in an uncertain atmosphere, in term of cost of power production, allotment and transmission while satisfying the power demand. To solve the model an algorithm was used based on Meta model-namely A-STRONG.

Tsuanyo et al. [12] proposed a model with variability in solar irradiation and the electrical loads to level the cost of energy. Dissimilar size of Ideal Diesel Generators (IDG) and Diesel Generators (DG), Conventional Diesel Generators (CDG) system is used for operation. Result confirmation is done by HOMAR software.

Yap et al. [13] used ANN for PV/diesel hybrid model and compare between Simulink model and an obtainable manufacturing model tool for a distant area in the northern Australia, data used for ANN training. Simulation results showed that the developed model is a viable planning and analytical tool for aiding future off-grid PV-to-diesel system integration applications, with R2 values ranging from 0.92 to 0.99 and mean relative errors below 5%.

Tanveer Ahmad [14] proposed dc-link which is an energy management scheme HRES fed by solar photovoltaic (PV) to reduce current stress and current limit on battery. DC linked used for voltage directive, successful power organization to set limit the current for both the battery and the supercapacitor. System compare with united cascaded control.

Mixture or different techniques for optimization can recover the limitation of particular methods. Some examples are: SA-Tabu search; Monte Carlo simulation (MCS)-PSO; hybrid iterative/ GA; MODO (Multi-objective design optimization)/GA; Artificial Fiber Fuzzy Interface Scheme (AFFIS); artificial neural network /GA/MCS; PSO/DE (differential evolution); evolutionary algorithms; and simulation optimization-MCS which often using in respective studies for optimizing of HRESs [15, 14].

Prakash Kumar et al. [15] used genetic algorithm for optimization process for MPPPT based solar system with integrated battery; and found the uninterrupted power for selected region, perturb and observe (PO) algorithm used for comparison.

Tzamalis et al. [16] examined independent power supply of rural and remote buildings with the help of Photovoltaic -diesel and Photovoltaic -hydrogen supply. He showed that the cost of energy (COE) of Photovoltaic -hydrogen system was higher than Photovoltaic -diesel system, but the reduction of greenhouse gas can be done by Photovoltaic-hydrogen. Moreover, the sensitivity analysis indicates that COE for the latter system can be further reduced as compared to its initial value by parameters like PEM electrolyzer and fuel cell capital costs.

Ghasemi et al. [17] presented a relative study between potential configurations of a system best suited to meet the isolated Iranian communities with the help of HOMER software. Author worked on NPC,

Non-Conventional fraction and air contaminant emission and showed that the stand-alone HRES composed of 15 kW PV array, a 20-kW diesel generator and a 20 kW inverter can supply 200 kW h/d energy consumption with a peak demand of 18 kW.

Ramli et al. [18] analyzed benefits of PV/diesel HES as energy sources with the flywheel as a storage system for Makah, Saudi Arabia using HOMAR software. The analysis focused on the impact of utilizing flywheel on power generation, energy cost, and net present cost for certain configurations of the hybrid system, and concludes that the power charge, and NPC, using up of fuel and C discharge can be reduced by using flywheel.

Zhou et al. [19] talked about merits and demerits of different optimization listed in Table 10.

Table 10. Advantages and disadvantages of different optimization technique

Software Tools	Advantages	Disadvantages
HOMER	It is freely available, and it can give efficient output.	Cannot enable the client to naturally select suitable design apparatus.
HOGA HYBRIDS	Genetic Algorithms (GA) can use in these software. Single as well as multi objectives can be used for Optimization. It requires very high knowledge for system operation.	It can simulate single design at one time.
Optimization Techniques	Can be use globally and it is suitable for complex parameters.	Its coding is very complex.
Graphic construction method Probabilistic approach Iterative technique	It doesn't need and data as time and series because any dynamic changing dies not affect the work.	It can include only two parameters for optimization.

Erdincand M Uzunoglu [60] compared dissimilar approach apply for the sizing of the HRES which is shown in Table 4.

Belmiliet al. [20] evaluated the Loss of Power Supply Probability (LPSP) algorithm for size optimization of HRES to satisfy the load profile.

Castaneda et al. [21] used different Simulink Design optimization (SDO) and control strategies like operating modes, several operational states, and modes of technical-economic operation for energy management of PV/hydrogen/battery HRES.

Table 11. Different approaches for the size optimization of HRES

Energy Organization	Pros	Cons
HOMER	It is freely available, and it can give efficient output	It cannot show characteristics the initial level linear equation base model
HYBRID 2	There are not any downloading charges for this software.	It utilizes "black box" code
Genetic Algorithm(GA)	It is suitable for complex parameters.	Its coding is very complex
Particle Swarm Optimization (PSO)	Easy coding	Low performance than GA, and not appropriate for difficult problem.
Simulated Annealing	Easy coding and literature reviews are also available easily	Low performance than GA, and not appropriate for difficult problem

Yang et al. [22] advised for a PV/WT with battery HRES in china, while Ekren et al. [23] showed same HRES for Turkey. Khan et al. [24].analyzes time series of a wind –fuel cell system. Onara et al. proposed wind/FC/ ultra-capacitor (UC) HRES and Veerachary [25] implemented PV model coupled with SEPIC converter to track the power of nonlinear PV sources.

Chauhan et al. [26] results that the under-sizing leads to failure of power supply while over sizing the system components will further add up the system cost.

Dufo-López et al. [27] has emphasized that if the number of variables and possible solutions are very high, the classic optimization techniques are failed to give appropriate best results and in those design problems heuristic techniques for optimization have been successfully applied. Among many, one of the generally used heuristic techniques is *Artificial Bee Colony Optimization*.

7. STUDY AREA

Uttarakhand, a state in northern India crossed by the Himalayas, has 53,483 km² land area, it consisting of 13 district viz. Almora, Bageshwar, Chamoli, Champawat, Dehradun, Haridwar, Nainital, Pauri Garhwal, Pithoragarh, Rudraprayag, Tehri Garhwal, Udham Singh Nagar, Uttarkashi in Garhwal Division. The district Tehri Garhwal, consist of 9 blocks, which are; Pratapnagar, Bhilangana, Jakridhar, Dhauladhar, Chamba, Narendranagar, Devprayag, Kirtinagar. Dudhli near Mussoorie was found appropriate remote rural area to be a subject for a techno-economic study on rural electrification by proposed hybrid energy system.

The "Queen of the Hills," Mussoorie is situated at an average altitude of 6950ft or 2005.5 mts from the sea level in the state of Uttrakhand, India. The beautiful city of Mussoorie is spread over an area of 64.25 kms in the Dehradun district of Uttrakhand. The geographical location of Mussoorie is 30.45°N 78.08°E.

The south of the hill station is occupied by the picturesque Doon Valley and the Shivalik Ranges that invigorates the beauty of the place. As far as population of the town is concerned, according to the census report of 2001, the population of the place was approx. 26,069. However, at present, Mussoorie is inhabited by approx. 34,000 people. Mussoorie receives an average rainfall of 180cms per annum.

The average literacy rate of Mussoorie is 79%. In addition to that, Mussoorie also serves as the Yamunotri and Gangotri, the most revered shrines in Northern India.

Mussoorie situated at 6950 ft. height from sea level in the state of uttarakhand.it is also as "Queen of the Hills. Dudhli situated about 15-16 km away on the way of Bhadraj Temple. Maximum land area occupied by agriculture. The geographical location of Mussoorie is 30.45°N 78.08°E. Condition of electricity is not very good (site visited on 2014 November) some villagers were using lantern in evening time. Geographical map of study area is showing in Figure 19.

Techno-Economic Analysis of Standalone Hybrid Energy System   115

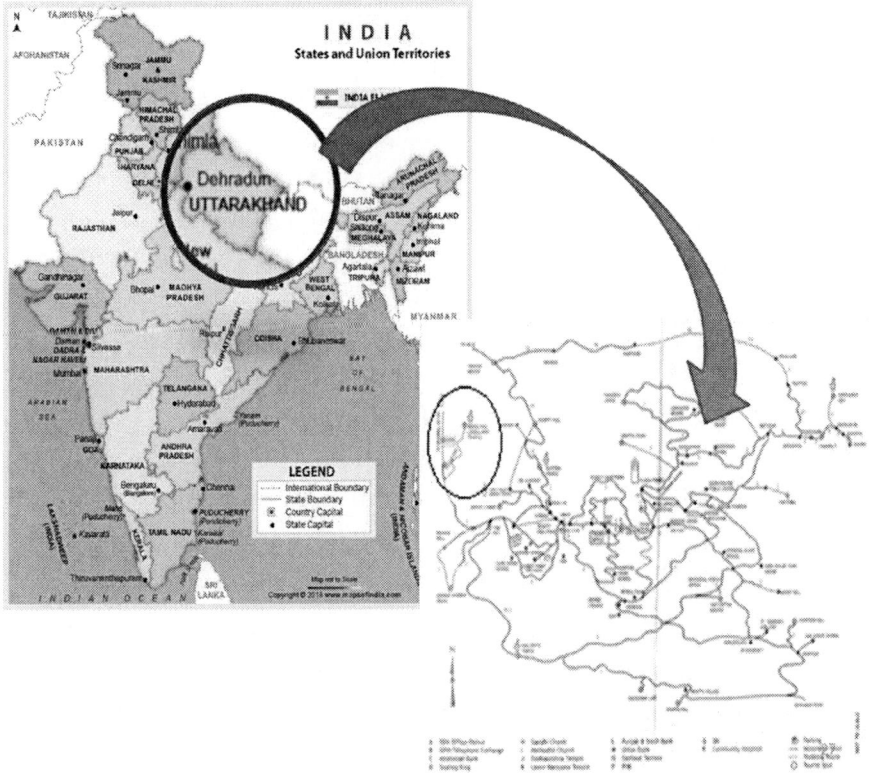

Figure 19. Geographical representation of selected site.

8. PHOTOVOLTAIC MODULE

Photovoltaic effect has been absorbed in 1883 by Edmund Becquerel. They saw a small voltage between the metals placed in semiconducting electrolyte when exposed in lights. Overall generation from it depends on several factors like type of material used in fabrication, type of connection used to form array, load forecasting, meteorological data of installation site, installation area requirement, etc. Solar PV generation is now a matured technology and depending on type of application, different types of solar cells can be used to convert solar radiation in to useful electrical energy.

8.1. Types of Solar Cells

Solar cells are made of semiconductor material, suitably doped by trivalent and pentavalent impurity to obtain p-type and n-type semiconductor respectively to manufacture solar PV cells. Solar PV cell designs are based on band gap energy (eV). Low band gap energy has high current ($I=eNA$) but low voltage ($V=E_g/e$), and vice versa. Here, e is electron charge, N is the number of photons, E_g is energy gap, and A is the surface area of solar cell. Therefore, it is preferred to use solar cell materials with energy gap between 1 and 1.8 eV like crystal silicon (1.12 eV), amorphous silicon (1.75 eV), copper indium diselenide (1.05 eV), cadmium telluride (1.45 eV), gallium arsenide (1.42 eV) and indium phosphate (1.34 eV). Among these, most commonly used material group is silicon cells. The silicon cell with highest efficiency is mono-crystalline cell, where the atoms are symmetrically placed within the structure. It gives high efficiency but involves high manufacturing cost of this symmetric structure. These modules are rated in terms of peak kilowatts (KW_p) i.e., the amount of expected electrical power output when the sun is directly overhead on a clear day. Commercial silicon cells have an efficiency of 14-17%. Polycrystalline cell is less expensive, where the structure is less symmetric and less complex. Its efficiency is about 13-15%. Another type of silicon solar cell is thin film cell based on amorphous silicon that is sprayed on a transparent surface. It differs from crystalline cells in terms of the amount of silicon needed and supporting structure required. For thin film cells a very thin layer of silicon is required and involves material like glass for holding the silicon together. Due to the low amount of silicon needed the required energy for manufacturing and thereby the production cost of those cells can be very low. The efficiency of the commercial thin film cells is about 5-7% with laboratories efficiencies of around 13%. Other materials involved in the thin film technology are copper, indium, gallium and selenide, for so-called CIGS cells, or cadmium and telluride, for CdTe cells. Toxicity of cadmium & telluride and the scarcity of indium have restricted their wide application.

8.2. Electricity Generation in PV Cell

PV generation is based on the principal of photoelectric effect. PV cell is fabricated with doped semiconductor p-n junction in the form of thin layer of semiconductor. When electromagnetic radiation (photon) with sufficiently high energy is incident on surface of solar cell, electron gets ejected and develops a potential between two semiconductor layers. This potential is observed as useful electrical energy, when connected across suitable load. Photon can get reflected, absorbed or pass through semiconductor; and the amount photons that can be trapped by PV cell determines its efficiency as shown in Figure 20.

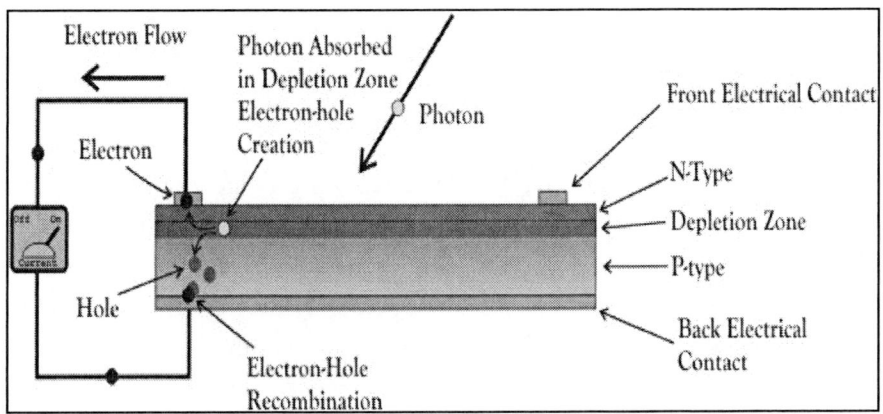

Source: Google Image.

Figure 20. Generation from PV cell.

Therefore, the fraction of photons that reflected should be condensed. An anti-reflective coating is typically useful at the outside of the Photovoltaic cell to minimize the total reflected photons. If a photon has lower energy than then energy of band gap of semiconductor, the photon is not capable to produce a pair of electron-hole and the semiconductor will not attract the photon. While if a photon has more energy than the band gap of the semiconductor the photon is absorbed by the valence band electron and any excess energy is emitted as a form of heat while the electron settles down in the conduction band. To reduce the percentage of photons that pass through, some semiconductors are manufactured with several layers, each having a different band gap to maximize the number of photons that are absorbed.

8.3. Factors Affecting the Power Generation

The angle of the incident light, irradiance, temperature, atmospheric condition, site location, clearness index and surface area of PV array are some of the factors affecting power generation capacity of PV module. The number of photons hitting a horizontal surface decreases with the angle from the perpendicular normal line according to the cosine law. Experiments and models show that the power generated by a solar cell decreases even more than expected by the cosine law, especially for angles >30° from the normal line. The reason is mainly that the reflection from the coating will increase with steeper incident angles. The decreased generation due to reflections is 11% and 50% for 60° and 80° respectively from the normal line. The temperature of a solar cell affects the shape of the current-voltage (I-V) curve. Higher temperature means increased internal energy losses and an increased band gap for the semiconducting material. This leads to a decreased V_{oc} and a slightly increased I_{sc}. The maximum power output is lower for a high temperature solar cell. The efficiency of the cell will decrease with about 0.4%/°C. An increased temperature also means higher stress on a panel, which decreases the lifetime by a factor of two for about every 10°C.

Solar array is the series-parallel combination of solar cell, connected to increase the output power. In an array consisting of series connected PV cells, a variation is observed due to mismatch in current generated in each cell. The cell producing the lowest current will decide the effect. This means that if one of the cells is shadowed the generation will drop for the whole panel due to the mismatch effects. The specified power for a PV panel is rated with perpendicular solar radiation of 1000 W/m² and a temperature of 25°C.

Other factors like dusty climate, snow logs, etc. can deteriorate power generation ability of PV module. For ideal generation, site of installation of PV panel should be in open region i.e., shading due to trees or buildings should be avoided. Even the shading of one PV cell can deteriorate overall generation capability up to more than 50%.

8.4. Technological Improvements to Optimize PV Generation

The generated current is directly proportional to the incoming radiation. So an increase in incoming radiation refers to higher output. The radiation can be reflected with mirrors or other material and focused to the solar cells. The increased output varies a lot between different designs and for each design it varies during the year due to radiation angle. One design is the Archimedes system, which is a solar tracing panel with V-formed concentrators. The system is made of solar cells covering 50% of the area and reflectors covering the other 50%. The enhancement of the effective radiation is about 1.5 -1.6 times, but varies depending on the weather conditions since only direct sunlight is reflected as modeled. The tracking system is increasing the output with another 1.25–1.35 times. To decrease the cell temperature, aluminum fins at the rear part of the panels are used. A simpler method is to use a fixed panel with a plane reflector. It can perform measurement of the annual generation. In this configuration, two panels are considered, one with and one without one plane reflector. The reflector has the same width (W) as the panel. The length of the reflector (L) is equal to the length of the panel plus the width on both sides of the panel ($L+2W$). The extended length is required to reflect radiation when the sun is not perpendicular to the attachment. The reflector is tilted twice a day and is modeled to reflect light on the whole panel during three hours before and three hours after noon. The cost of the reflector is about 5% of the cost of the panel. The power output increases over the year with about 22%. The negative impact of the reflector is an increased cell temperature. The surface temperature increases with about 10°C, compared with the panel without reflector. Another important way to increase the power output and make the investment cost lower for a stand-alone system is to make sure that the power supply is equal to the power consumption. If the consumption gets lower than the estimated value, the investment will be higher than needed.

8.4.1. Performance PV Module

Figure 21 shows the equivalent circuit of PV cell which consists a photodiode, 1 series and parallel resistor. According to KCL(Kirchhoff's current law) PV current defined by eq.1. where I_{gc} generated current I_o is the diffusion current, e is the electric charge which is 1.6×10^{-19} Coulombs, K is Boltzmann's constant = 1.38×10^{-23} J/K, F is the cell idealizing factor, T_c is the total temperature of cell, v_d= diode voltage, and R_p= parallel resistance.

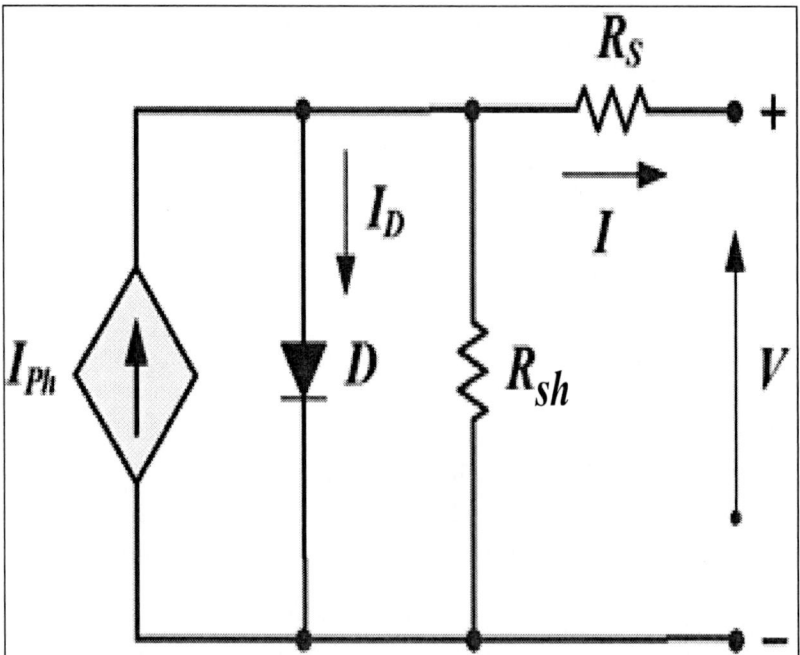

Figure 21. Diode model of PV cell.

The performance of PV modules can be analyzed by performing mathematical model of current-voltage (*I-V*) characteristic of PV modules for given meteorological data. Ideally, it is performed on the PV module represented by a diode model as shown in Figure 21.

The model includes a constant current source representing PV cell, a diode, a shunt resistance R_{sh} and a series resistance R_s. The basic terms involved in PV conversion are given as:

$$I = I_{ph} - I_D - \frac{V_D}{R_{sh}} = I_C - I_D - I_{R_{sh}} \tag{20}$$

$$V_{PV} = V_D - R_s I_{PV} \tag{21}$$

$$I_D = I_0 \left(e^{(V_D/akT)} - 1 \right) \tag{22}$$

$$T = T_a + \left(\frac{NOCT - 20^0}{0.8} \right) S_i \tag{23}$$

Where, I_0 is saturation current, V_D is diode voltage, I_D is diode current, k is Boltzmann's constant (1.38 x 10^{-23} J/K), a is numerical constant (1 for germanium and 2 for silicon), T is junction temperature, T_a is ambient temperature, and $NOCT$ is cell temperature in a module when ambient temperature is 20°C. Change in the characteristics due to the variation in solar insulations and temperature is shown in Figure 22.

$$\eta = \frac{P_{mo}}{P_{in}} = \frac{P_{mo}}{AG_i} = \frac{(V_{oc} . I_{sc}) FF}{AG_i} \tag{24}$$

Where, $V_{oc} = \frac{akT}{e} \left[\log \left(\frac{I_L + I_0}{I_0} \right) \right] \tag{25}$

$$I_{sc} = I_{PV} - I_0 \left[\exp \left(\frac{eR_s I_{sc}}{akT} \right) - 1 \right] - \left(\frac{R_s I_{sc}}{R_{sh}} \right) \tag{26}$$

And, $FF = \dfrac{V_m \cdot I_m}{V_{oc} \cdot I_{sc}} = \dfrac{P_m}{V_{oc} \cdot I_{sc}}$ (27)

Here, G_i is solar irradiance, V_{oc} is open circuit voltage, V_{sc} is short circuit voltage, and FF is fill factor (0.5 to 1).

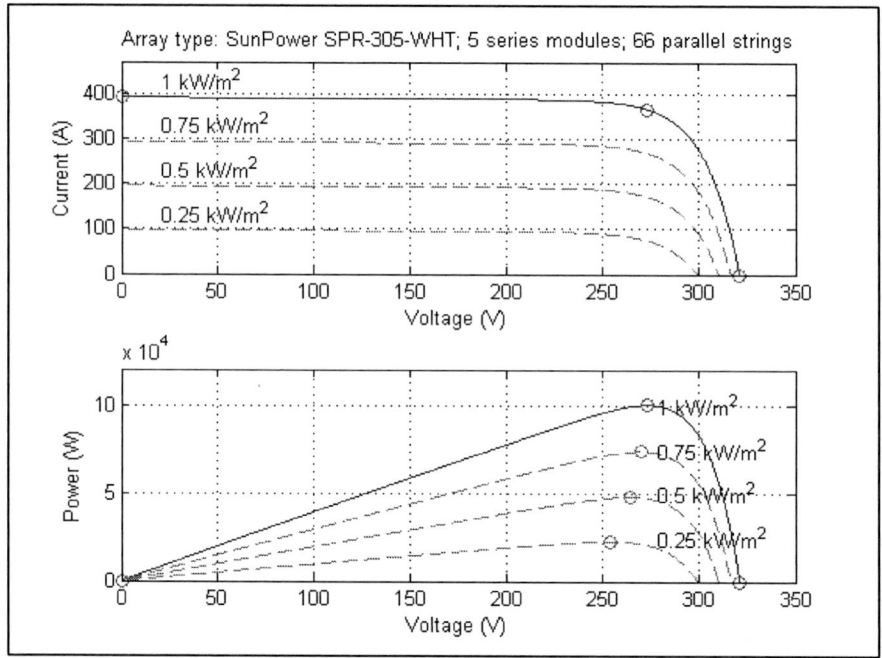

Figure 22. Change in the characteristics due to the change in temperature. (a) Current (b) Power.

Figure 23 shows the relationship between current and voltage at different Irradiance. It can be seen that as the irradiance varies the current and voltage also varies. The current increases with the voltage. After a saturation point (known as maximum power point) the current starts decreasing with the further increase in voltage. The maximum power point is the point at which voltage and power are maximum for a particular irradiance level. In making the graph, the condition is considered as an ideal ignoring the environmental aspects.

The second graph depicts the relationship between voltage and power. Initially with the increase in voltage, the power increases proportionally but after the MPPT point (saturation point), the power starts decreasing with increase in voltage. So, after a particular voltage, irrespective of irradiance the current and power become zero.

8.4.2. I-V Characteristics

By using single diode, as shown in Figure 23, equivalent electric circuit of PV cell has been used to obtain current-voltage (*I-V*) characteristics of a PV cell. The mathematical model of a photovoltaic cell can be developed using Matlab/Simulink toolbox. The basic equation from the theory of semiconductors that mathematically describes the *I-V* characteristic of the ideal photovoltaic cell is given by:

$$I = I_{pv,cell} - I_D \tag{28}$$

$$I_D = I_{0,cell}\left[\exp\left(\frac{qv}{akT}\right) - 1\right] \tag{29}$$

Therefore

$$I = I_{pv,cell} - I_{0,cell}\left[\exp\left(\frac{qv}{akT}\right) - 1\right] \tag{30}$$

Where, $I_{pv,cell}$ is the current generated by the incident light (it is directly proportional to the solar irradiation), I_d is the diode current, $I_{0,cell}$ is the reverse saturation or leakage current of the diode, q is the electron charge [1.60217646 x 10^{-19}C], k is the Boltzmann constant [1.3806503 x 10^{-23}J/K], T is the temperature of the *p-n* junction, and a is the diode ideal constant.

Practical arrays are composed of several connected PV cells and the observation of the characteristics at the terminals of the PV array requires the inclusion of additional parameters to the basic equation:

$$I = I_{pv} - I_0 \left[\exp\left(\frac{V+IR_S}{V_t a}\right) - 1 \right] - \left[\frac{V+IR_S}{R_{sh}} \right] \tag{31}$$

Where $V_t = N_s kT/q$ is the thermal voltage of the array with N_s cells are connected in series. Cells connected in parallel increases the current and cells connected in series provide greater output voltages. V and I are terminal voltage and current. Typical I-V characteristic of a solar cell in steady-state operation has been shown in Figure 23.

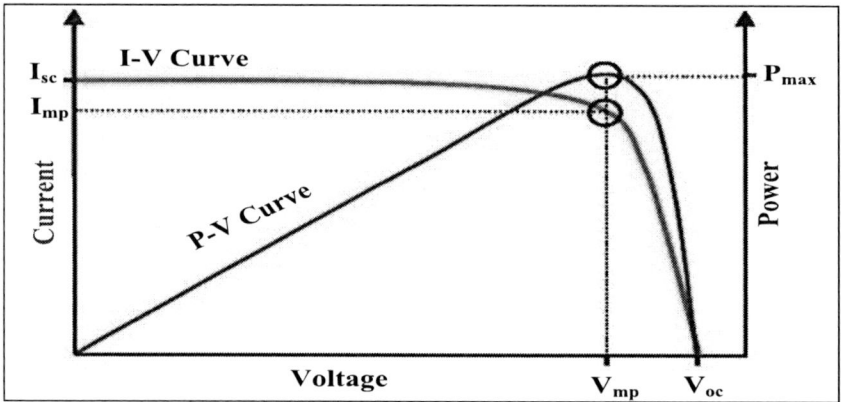

Figure 23. Typical I-V characteristic of a solar cell in steady-state operation.

Table 12 shows the input parameter rating for a single module and total 6 modules are in parallel and 10 in series taken for this work.

Table 12. Solar input parameter for single module

Parameters	Rating for single module	Rating for total used module 6 parallel and 10 series
Rated power	225W	13500W
Highest power voltage	29.76V	1785.6V
Highest power current	7.55 A	453A
Open circuit voltage	36.88 V	2212.8V
Short circuit current	8.27 A	496.2A
PV cell used in sequence	16	
PV cell used in parallel	13	

8.5. Designing of Wind Turbine

A wind energy system is generally employed where the speed of wind is more than 10m/s. Thus to describe the operation of Wind Turbine system, a MATLAB block of WT is used. In the block the general equation of mechanical power(P_{mec}) generated is given as

$$P_m = C_P(\lambda, \beta) \frac{\rho A}{2} V_{wind}^3 \tag{32}$$

Where ρ=Air density
A=Area swept by blades
Cp=Coefficient parameters which are constant
λ, β=Tip ratio and pitch angle
Tip Ratio=Velocity of wind striking the tip of blade/Total velocity of the wind

Temperature coefficient which are used in this paper are taken form [12]

$$C_P(\lambda, \beta) = C_1 \left(\frac{C_2}{\lambda_i} - C_3\beta - C_4\right) e^{\left(\frac{-C_5}{\lambda_i}\right)} + C_6\lambda \tag{33}$$

Parameters of C_1 to C_6 depends on rotor of WT and design of blade and λ_i given by

$$\frac{1}{\lambda_i} = \frac{1}{\lambda + 0.08\beta} - \frac{0.035}{\beta^3 + 1} \tag{34}$$

Now per unit power (Pm-pu) is

$$Pm - pu = K_p C_{p-pu} V_{wind-pu}^3 \tag{35}$$

Where
c_{p-pu}= performance coefficient

kp = energy gain

$v_{wind\text{-}pu}$ = speed of wind m/s

Simulink model of WT shown in Figure 2 in Appendix A Sim power system library is used to build WT where generator, pitch angle, wind speed has been used as the input of WT and the output will be torque applied on the WT shaft.

Table 13 shows the input parameter used for this work. Generator speed calculated in different times during day and night time separately.

Table 13. Input parameter for WT

Generator speed	1 pu (Base Generator Speed)
Pitch angle	0
Wind speed at different times	[12 0 1 2 15]- during night and after sunset

8.6. Designing of Biomass Plant

The biodiesel plant modeled in this project is basically a photo bioreactor which means that it would need solar irradiance as one of its input for the production [14]. The biomass generated is considered as an initial (P_{mec}) for the synchronous machine. Again, the three phase AC is converted to DC using the diode rectifier.

The generalized logistic growth equation used to model the dynamics of the microalgae growth within the PBR

$$\frac{dB}{dt} = \mu B \left(1 - \frac{B}{K}\right) \tag{36}$$

Where, 'B' = Biomass inhabitants,

'K'= Carrying capacity of the space in which the biomass is harvested

'µ'=Maximum growth rate

Now

$$\mu = K_{PAR} I_{PAR} \tag{37}$$

Where, K_{PAR} = Sun parameter
I_{PAR} = Active radiation
Figure 3 in Appendix A shows the simulated model of biodiesel plant.

Table 14. Input data of biodiesel plant

Carrying capacity(k)	252g/m^2
Inhibition factor(θ)	5.6694
Death rate	0.0531/day
Growth rate	0.381/day

8.7. Boost Convertor

The output voltage generated by the solar PV array is not enough to run a three-phaseload. Boost converter used to increase the Voltage up to 450V to maintain the 400-415V for DC bus. Since a boost convertor involves a switch (say IGBT/GTO), the triggering/pulses of boost converter is provided through MPPT. The MPPT (maximum power point tracking) tracks the direction of sun at which power is maximum, so that maximum output can be extracted from the solar.

8.8. Battery Connections

A battery is connected in parallel with the overall hybrid system so as a back in the case of emergency, when none of the system works especially during the night time, although the standard battery available is generally 12/24V, but to avoid number of batteries in series or parallel, a single battery block available in MATLAB is used. The ratings are considered with respect to load. The initial charging state has been considered as 90% which would charge/discharge slowly and can be seen in SOC (State of Charge) graph. The battery can be charged with any of the source and will discharge if the DC bus voltage drops below the certain level. A bidirectional DC-DC converter has been used to charge and discharge the battery as per the required conditions.

8.9. Load Side Connection

A three phase 30kW resistive load is considered in the model. Since we are getting the DC voltage form PV array, so this DC bus voltage is converted to AC using PWM inverter. The output from the inverter is filtered using the LC filter. The PWM is synchronized with the load voltage to maintain the modulation index. All the respective waveforms are shown in different scopes. Complete simulated model is given in appendix A.

9. MPPT Techniques for Solar PV Generation

PV systems normally use MPPT technique to continuously deliver the highest possible power to the load when variations in the isolation and temperature occur. PV generation is becoming increasingly important as a renewable source, as it offers many advantages like incurring no fuel costs, non-polluting nature, little maintenance requirement, and non-involvement of ant noisy component. PV modules still have relatively low conversion efficiency; therefore, controlling MPPT for the solar array is essential in a PV system. MPPT is a technique used in power electronic circuits to extract maximum energy from the PV Systems. In the recent days, PV power generation has gained more importance due its numerous advantages such as fuel free, requires very little maintenance and environmental benefits. To improve the energy efficiency, it is important to operate PV system always at its maximum power point. Several MPPT techniques are available and involve variety of methods for obtaining maximum power point. The noteworthy methods reported in literature section 2.1.2.

The objective of MPPT algorithm is to adjust the current (I_{mpp}) and voltage (V_{mpp}) of the PV array at which maximum output power (P_{mpp}) is obtained under a specific irradiation and temperature. As mentioned earlier the P&O and INC are the most popular techniques for tracking of maximum power. Both P&O and INC are based on hill climbing algorithm. In this study, P&O based MPPT technique has been used to perform comparative study for proposed ABC algorithm based MPPT technique.

10. PERTURB AND OBSERVE ALGORITHM

P&O algorithm is very popular and the most commonly used in practice due to its simplicity and the ease of implementation. In this method the operating voltage of the PV module is perturbed by a small increment and the resulting change in power ΔP, is observed. If the ΔP is positive, then it is supposed that it has moved the operating point closer to the MPP. If the ΔP is negative, the operating point has moved away from the MPP and the direction of perturbation should be reversed to move back toward the MPP. Figure 24 shows the flowchart of this method.

In this work, P&O algorithm reads the value of current and voltage from PV module; and calculates power. In order to obtain local maximum power point, the value of voltage and power at k^{th} instant are perturbed and verified with next values of voltage and current at $(k+1)^{th}$ instant. During this observation procedure, power and voltage at k^{th} instant are subtracted from the values at $(k+1)^{th}$ instant. Now, it is observed from the power voltage curve of PV module that, power voltage is negative ($dP/dV<0$) in right hand side curve where the voltage curve has almost constant slope and in the left-hand side the voltage curve slope ($dP/dV>0$) is positive. The right-side curve refers to lower duty cycle (nearer to zero), whereas the left side curve refers to higher duty cycle (nearer to unity). Depending on the sign of $dP(P(k+1) - P(k))$ and $dV(V(k+1) - V(k))$; the P&O algorithm decides whether to increase the duty cycle or to reduce the duty cycle. Beside its better exploitation, it lacks exploration of optimal solution as it searches for local maxima and lacks large search space to explore global maxima or minima.

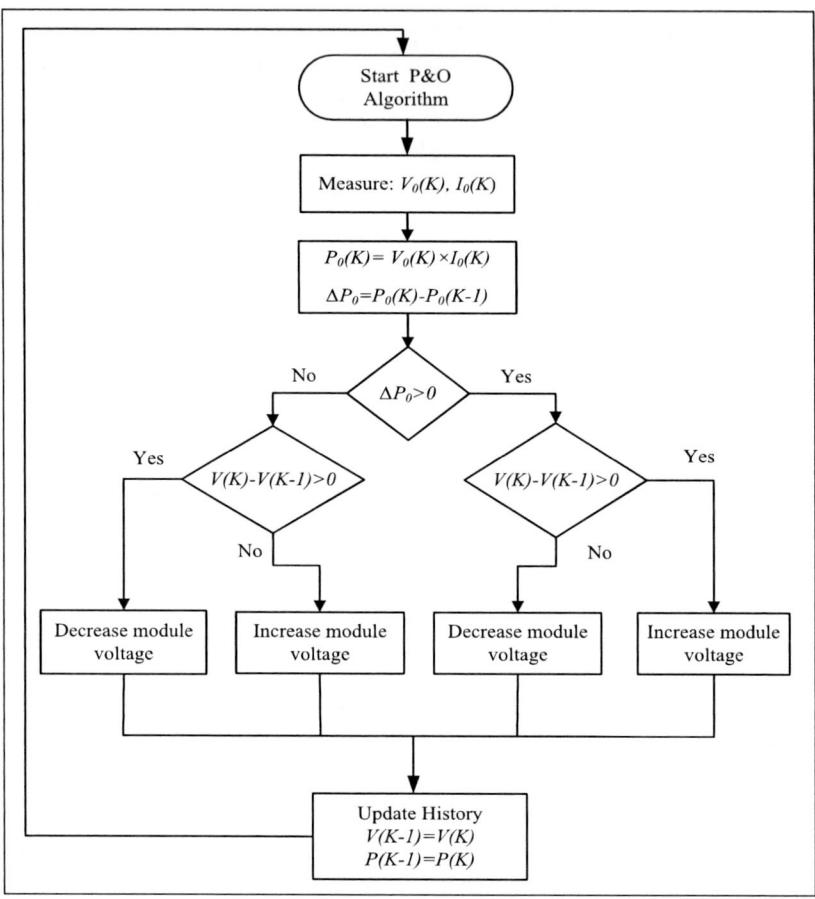

Figure 24. Flowchart of the P&O method.

11. ABC Algorithm

Karaboga et al. [94] introduced by a new evolutionary heuristic technique, known as artificial bee colony (ABC). ABC algorithm is formulated by noticing the behavior of real honey bees on discovering nectar at a site and sharing the information of the available food sources or nectar to the other bees in their hive. The preferences of Artificial Bee Colony (ABC) algorithm over other optimization techniques are:

- Simplicity
- High flexibility
- Strong robustness

Karaboga et al. [95] introduces few control parameters , and ability to optimize the objective function with fast convergence, stochastic nature and both exploration and exploitation by evaluating conditions in minimizing the total net present cost, considering the finest parameters for optimum configuration with high performance which implicates deciding on the mix of above parameters that the system must contain, the approximate number of each component, and the overall minimized cost of HRES system. A. Singh [96] explained ease of combination with other methods.

The colony of artificial honey bees is split into three different groups of bees which include active/employed bees, onlookers, and explorer/scouts. Each type of bees plays a different role in the overall optimization process. The active/employed bees stay around the nectar source and keep the neighboring position in memory while the onlooker bees accumulate that data from the active bees and make a suitable choice to choose a site and collect the nectar and the scouts are very much accountable for determining the new source position within the same area. The algorithm for solving an optimization problem has been comprised with 3 steps. In the step one, the employed bees are directed by queen bee to hunt for favorable resources and the nectar amount is measured. In the step two, the observer/onlooker bees make a suitable resource choice by observing acts of active bees on hive appropriate to the information they took from the newly identified nectar position. Finally, in step third, one of the active/employed bee is chosen at random as an explorer/scout bee and is sent to the sources to identify new sources position. Half populations of the bees in the swarm are chosen as active/employed bees and the rest halves are chosen as the observer/onlooker bees in the algorithm. Therefore, the number of the active bees is equal to the number of food sources considered. The food sources in the approach accredit to the probable solutions to the parameters to be optimized.

11.1. Steps for ABC Algorithm

To begin with the Artificial Bee Colony algorithm in the first step, the initial solutions are randomly chosen in the particular limits of the variables $x_i (i=1,2,3....S)$. Secondly, each active/employed bee makes out the neighborhood position whose numbers are equal to the half of the total available sources. (38) is used to find a new source as follows:

$$V_{ij} = x_{ij} + \Phi_{ij}(x_{ij} - x_{kj}) \tag{38}$$

In (1),k∈1,2,....N and j∈1,2,....D are random selected indexes as per the no. of parameters to be optimized. Though k is randomly considered, it has to be different from i and Φ_{ij} and is a random no. between 0 and 1. The parameter Φij controls the elongation of the neighbor food sources around and epitomize the comparison of two food source positions visually by bee. After each candidate source position is made and then evaluated by the artificial bee, its performance is correlated with its preceding one and accordingly which so ever food source has better or equal nectar than the old source; it is used to substitute the old one in the memory. Otherwise, the old one is with held in the memory.

In last step, the onlooker/observer bees select a food source with the maximum probability

$$P_i = \frac{fit_i}{\sum_{j=1}^{S_n} fit_j} \tag{39}$$

Where fit_i is the fitness value of the solution i which is relative to the nectar quantity of the food source in the location i and j is the no. of food sources which is equivalent to the number of active/employed bees.

The scout/explorer bees are very much liable for random lookout in each colony. The observer bees do not use any prerequisite-facts and knowledge when they are gazing for new nectar sources, thus their exploration is randomly done totally. The scout/observer bees are chosen among the employed/active bees with respect to the limit parameter. If a result that denotes a source is not realized within particular no. of trials, then that origin is discarded. The bee of that origin hunts for the new source and is chosen as a scout bee. The no. of incoming and outgoing to a particular source is obtained by the "limit" parameter. Identifying a new source for a scout/observer bee is given in (40).

$$X_{ij} = X_j^{min} + \left(X_j^{max} - X_j^{min}\right) * rand(0,1) \tag{40}$$

Where X_j^{min} and X_j^{max} are the minima and maximum limits of the parameters to be optimized.

In the ABC algorithm, the stopping criterion is usually taken on the maximum number of iterations. Normally, stopping criteria of an optimization algorithm is based on a maximum number of iterations or maximum error between two consecutive iterations. However, the maximum error cannot be applicable for met heuristic search methods since they usually suffer unimproved result for several iterations during the convergence process. Therefore, the stopping criterion for the proposed ABC is only based on the maximum number of iterations.

11.2. Implementation of ABC for the Hybrid System

The overall implementation of the ABC algorithm for this hybrid system optimization is as follow, flow chart of the algorithm is shown in Figure 24.

Step 1. Initialize random 50 to 100 for each variable such number of collectors, a diameter of the parabolic reflector, number of modules in series and gross area available for solar and biomass system. In the same way for wind turbine system the hub height, scale factor, diameter of rotor and coefficient of ground friction are initial food origin and parameters for the ABC algorithm such as limit, the maximum number of iterations etc.

Step 2. Calculate fitness function i.e., total cost for all the individual hybrid system consisting of solar PV, wind turbine, and biomass system. The overall cost of any system consists of initial cost of individual components, operation and maintenance cost, salvation cost and subsidy provided by a government.

$$C_T = C_I + C_M + C_R - C_s \qquad (41)$$

Where C_I is the initial cost, C_M is the operation and maintenance cost, C_R is the salvation cost and C_s is the subsidy

Step 3. Initialized the chosen parameters as the best solution. Set the iteration counter to 1.

Step 4. Actuate the position for active/employed bees which is basically the parameters chosen in our model.

Step 5. Calculate the fitness function (costs) based on the location of active bees.

Step 6. Recalculate the fitness function value (costs) corresponding to the new location of the active/employed bees.

Step 7. Compare the new obtained values of the fitness function(costs) with the initial (costs) to pick up the best one by considering the minimum between the two.

Step 8. If not all observer/onlooker bees are distributed to food origins, update the new location for the observer bees and return to Step5.

Step 9. Determine unchanged food sources when the "limit" exceeds.

Step 10. If any unchanged food sources were found, initialize that food source for the observer's bees, and calculate fitness function value.

Step 11. Update the best food source location and corresponding fitness function values.

Step 12. If the maximum no. of iterations is not reached, increase the iteration counter and return to Step 4.

Step 13. The final optimized values obtained are used in the Simulink model in the following given equation which is the model by using Simulink blocks.

Techno-Economic Analysis of Standalone Hybrid Energy System 135

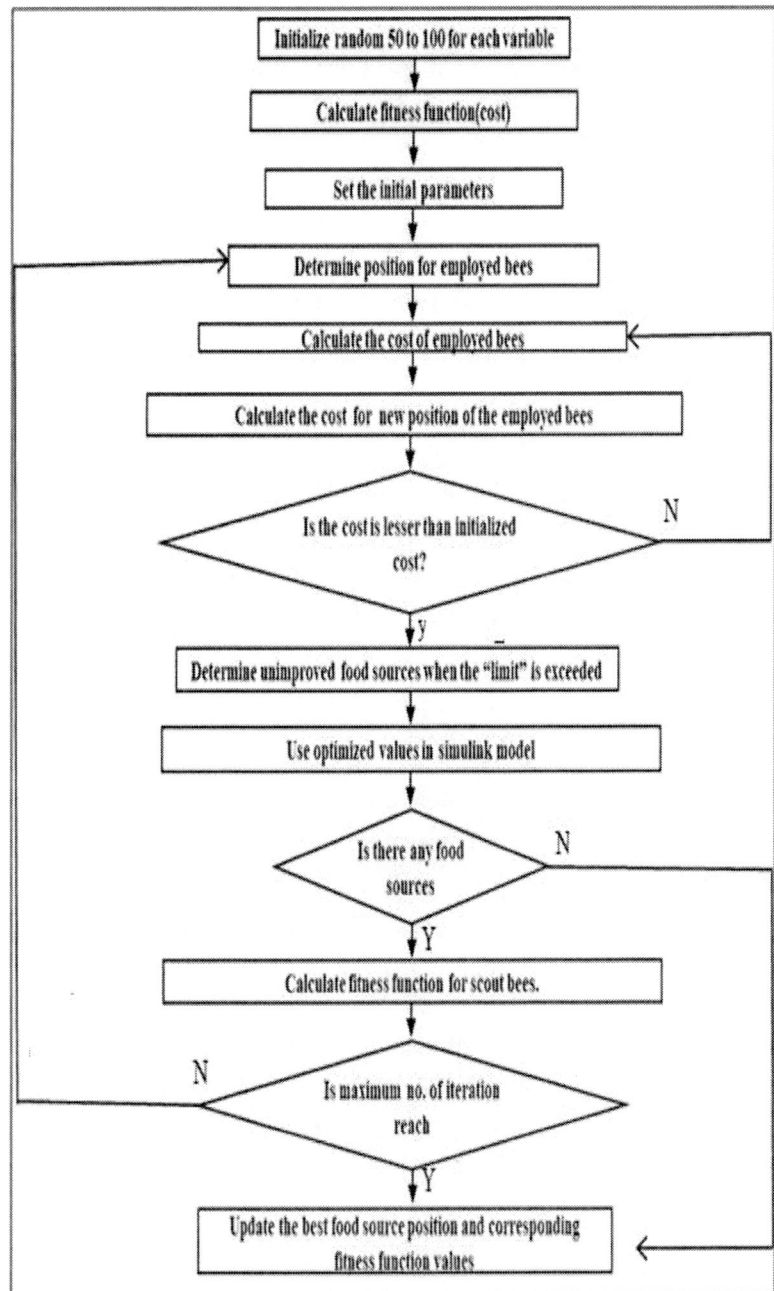

Figure 25. Flow chart for ABC implementation.

The Weibull probability density function is mainly used for calculation of AEP because of its synoptic nature and the tendency to demonstrate the random variation of the wind speed. The equation modeled is given as

$$W(v) = \left[\left(\frac{m}{c}\right)\left(\frac{v}{c}\right)^{m-1} \exp\left[-\left(\frac{v}{c}\right)^m\right]\right] \qquad (42)$$

Where v denotes steady wind speed. The environment conditions also influence the wind distribution; the parameters m and c are the function of hub height and are defined as:

$$C = C_0 \left(\frac{h}{h_0}\right)^\alpha \qquad (43)$$

$$m(h) = m_0 + \Delta m(h) \qquad (44)$$

$$\Delta m(h) = \begin{cases} 0.008 - 0.08 \text{ for } h < 20m \\ 0.003h + 0.02 \text{ for } h \geq 20m \end{cases}$$

Where h is the hub height (in meter), c_o is the value of c at the reference height(h_o), and α is the coefficient of the ground surface friction.

The degree of optical (short wavelength) radiation that incidents on absorber/receiver will be the solar irradiance resource for that type of collector and its tracking (global (total) solar irradiance for a flat-plate collectors and direct (beam) solar irradiance for a concentrating type collector). Since the capture area of the collector may not be intended directly at the sun, this resource must be reduced to account for the angle of incidence. The area of the collector on which the solar irradiance falls is called the opening (aperture) area of the collector.

$$A = \frac{2}{3} * (b * h) \qquad (45)$$

where b is a length of the chord crossing the focus and is perpendicular to an axis of symmetry

h is a length of the perpendicular distance from the parabola's vertex to the chord

The focal length, f, of the parabola is given by:

$$f = \frac{b^2}{(16*h)} \tag{46}$$

The Weibull probability density function and the parabolic reflector are modeled in the model and the optimized values of respective parameters are used in it.

12. COST COMPARISON GLOBALLY

Major Cost function of a HRES included initial cost, operation and maintenance cost yearly, salvation costs and recovery cost or salvages values of each component. Table 15 shows the unit cost of different resources. Objective function can be formulated as:

Table 15. Unit cost of energy for different resources

S.No.	Types of energy resources	Cost of energy (Rs/Kwh)
1	Large hydro	1.31-12.42
2	Small hydro	1.31-17.64
3	c-Si PV system	16.34-42.47
4	c-Si PV system with battery storage	23.52-46.40
5	Biomass generator	4.57-16.99
6	Diesel generator	11.12-13.42
7	Battery	3.1-4.1

13. RESULT AND ANALYSIS

Based on the calculations, the developed algorithm simulates the hybrid energy system for one set of twenty hours' data (site load, solar potentials and scheduling of renewable generators) are gathered in each month from January to December shown it graphs below:

Figure 27 (A) show the actual wind speed before optimization with time. The wind speed is higher at night compared to the day time. (B) Shows the variation of optimized wind speed with time. It can be observed that wind speed is higher in night as compare to day time. For the modelling purpose we have considered one day weather as moderate. So, with the higher rate of wind speed most of the load requirements can be met by wind turbine plant. For improving the wind capacity a Weibull equation is implemented in the model after optimization the hub height, rotor diameter, scaling factor.

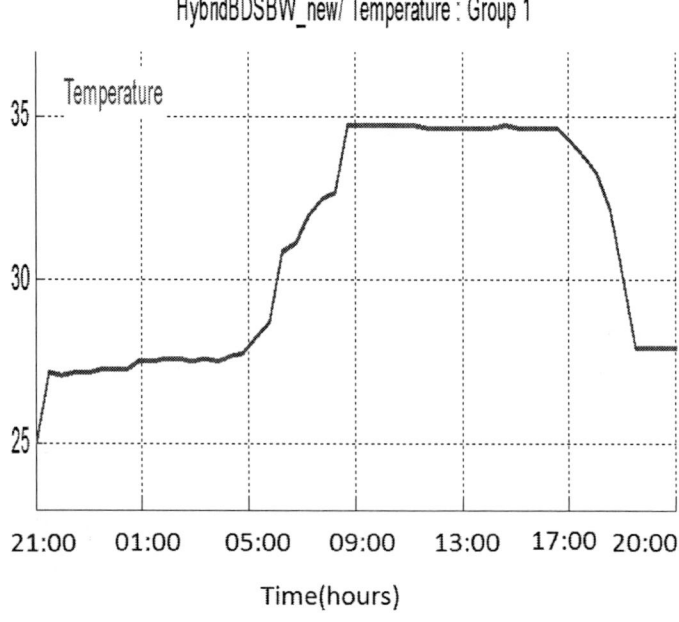

Figure 26. Input temperature of solar model.

Figure 28 A and B shows the actual and the optimized irradiance. using the optimizing technique, the new efficiency of irradiance level is improved up to 40%. This efficiency also boosts the system efficiency thus maximum power can be extracted just adding some extra components in the system. In this with the usage of parabolic reflector and optimization it can be observed how the overall output improves.

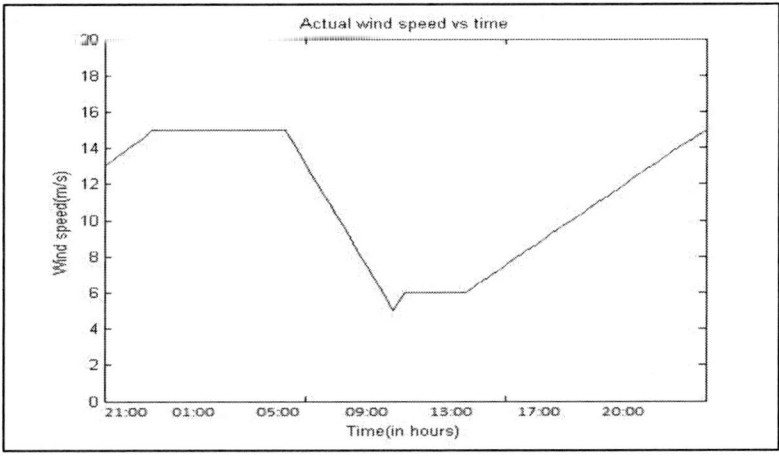

Figure 27. (A). Actual wind speed graph.

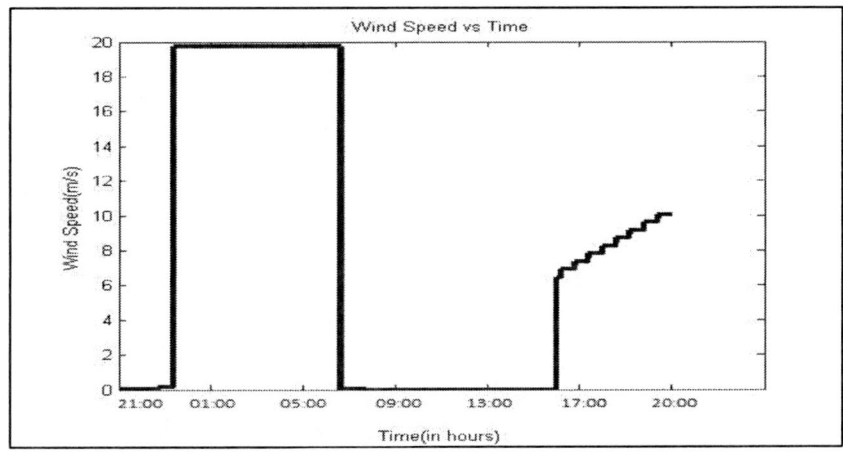

Figure 27. (B). Optimized wind speed graph.

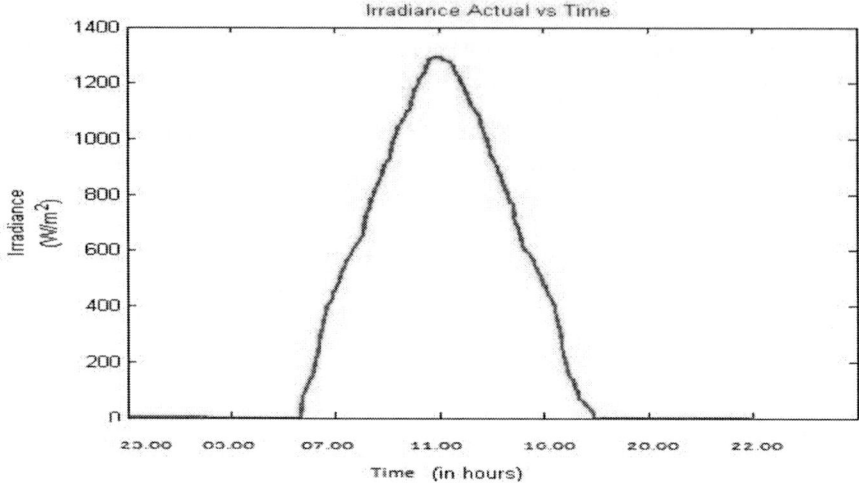

Figure 28. (A). Actual solar irradiation graph.

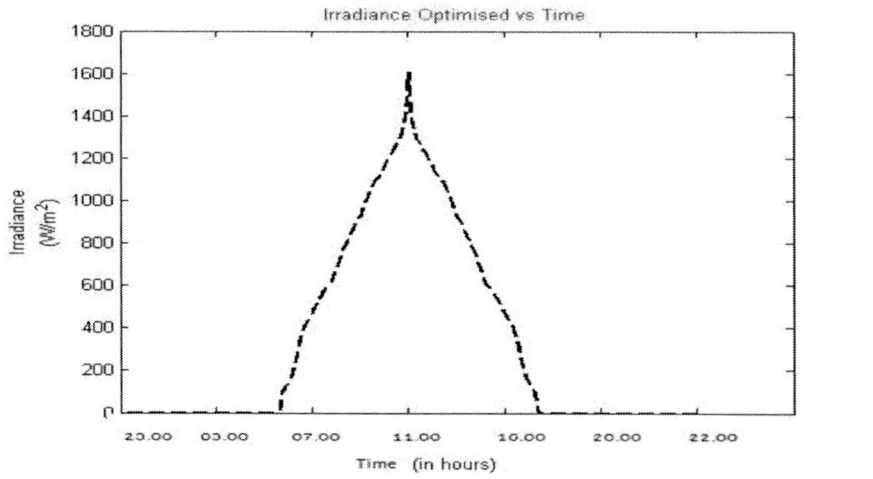

Figure 28. (B). Optimized solar irradiation graph.

Figure 29 switching diagram clearly shows the switching of each individual sources for the 24 hours duration of time. At the time when all the three sources are inactive, the battery as a backup fulfilled the required demand, thus maintaining the constant power flow.

Figure 30 shows the charging/discharging of battery with time. The battery charges when there is enough power left after providing to the load and discharges whenever all the sources are absent. It can be observed that during the day time, the battery charges from solar PV module and during night it charges with wind system.

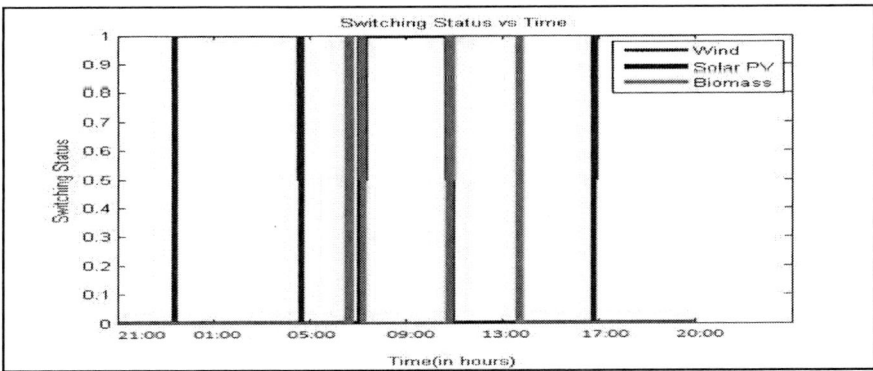

Figure 29. Different switching status.

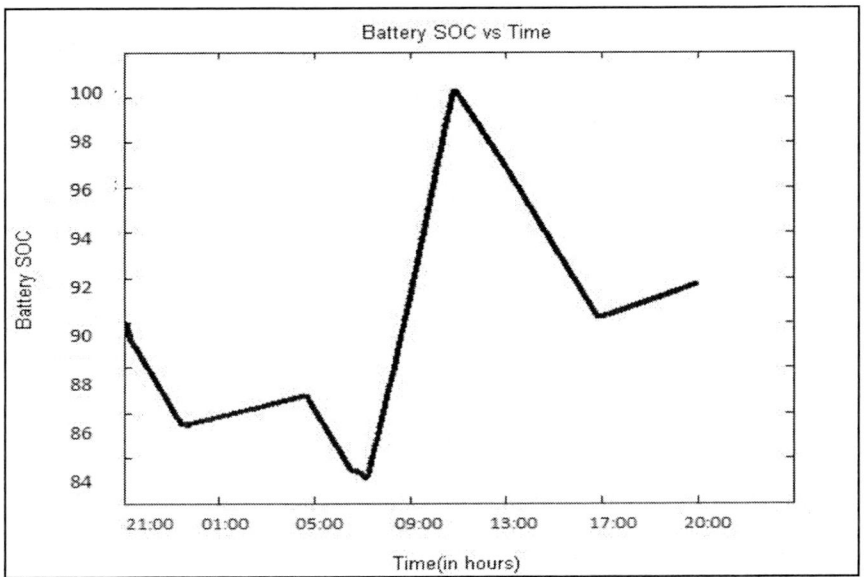

Figure 30. Battery charging.

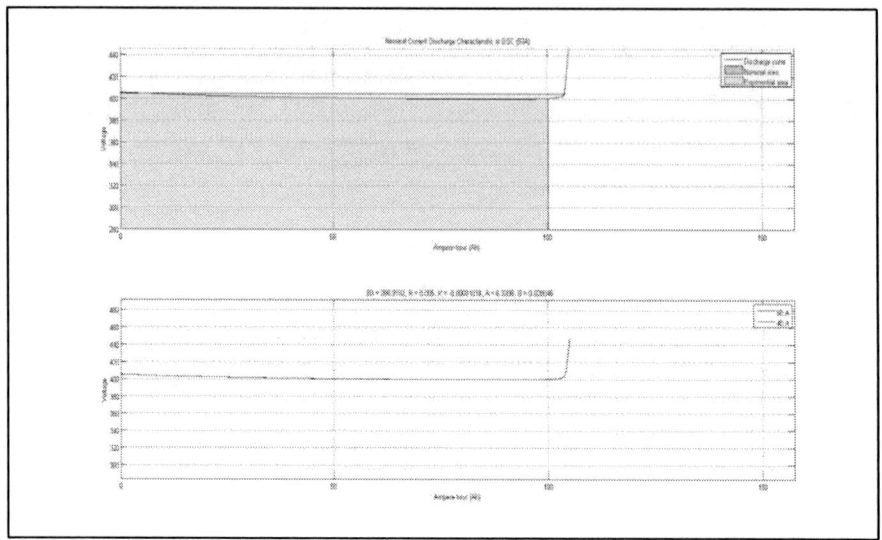

Figure 31. Battery discharging.

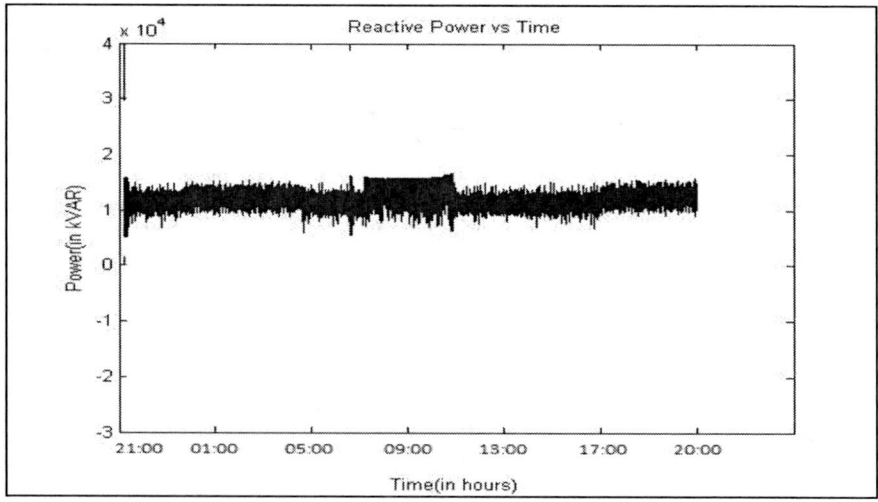

Figure 32. (A). Active power output of the load.

A typical discharge curve is composed of three sections, as shown in the Figure 31.

Techno-Economic Analysis of Standalone Hybrid Energy System 143

The first section represents the exponential voltage drop when the battery is charged. Depending on the battery type, this area is more or less wide. The second section represents the charge that can be extracted from the battery until the voltage drops below the battery nominal voltage. Finally, the third section represents the total discharge of the battery, when the voltage drops rapidly.

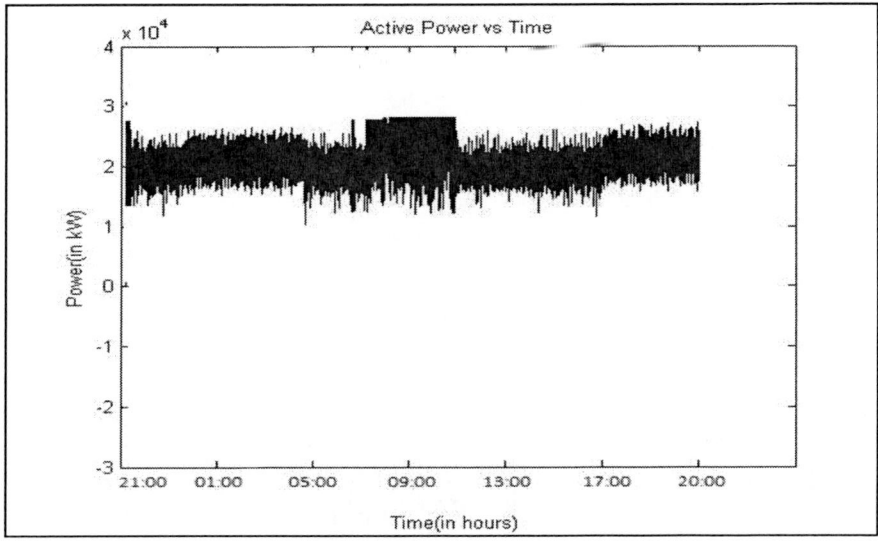

Figure 32. (B). Reactive power output of the load.

Figure 32 The load considered in modeling consist of both active and reactive power components. The above graph shows the variation of reactive power with time. It can be seen that a minimum reactive power of 18KVAR is available throughout the day and thus this can be used for inductive/capacitive load.

Figure 34 shows the variation of DC voltage across the inverter. It can be seen that DC voltage is always close to 400V, for all the different sources. As there is switching from one source to another there is slight variation in the DC output voltage, but it gets stabilized within minutes of operation.

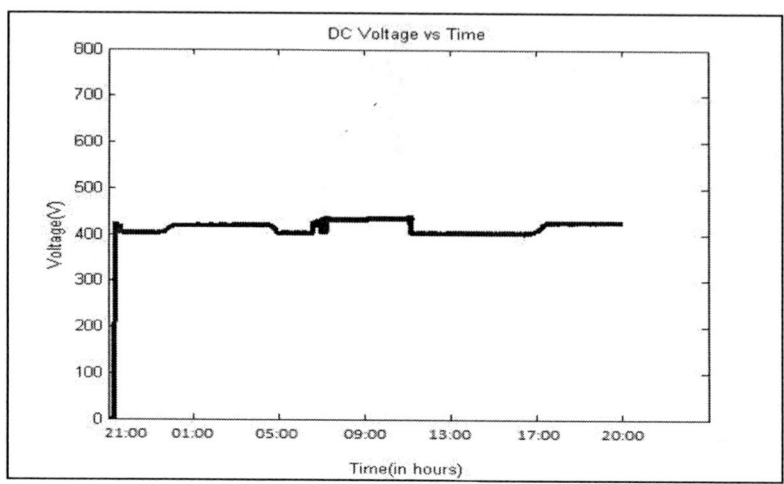

Figure 33. DC bus voltage.

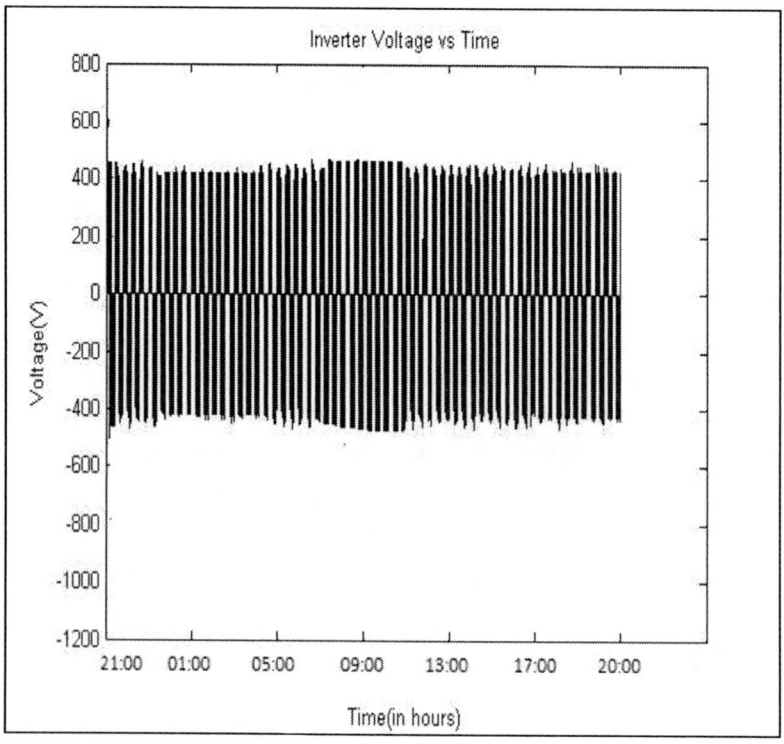

Figure 34. Inverter output voltage.

Figure 35 and Figure 28 (A and B) depict the relation between voltage and current across the load with respect to time. The graphs show that irrespective of source, the voltage and current across the load is maintained constant. The voltage is 400-415V (3 phase AC) and current is 40A. The constant voltage and current are obtained after using LC series across the inverter.

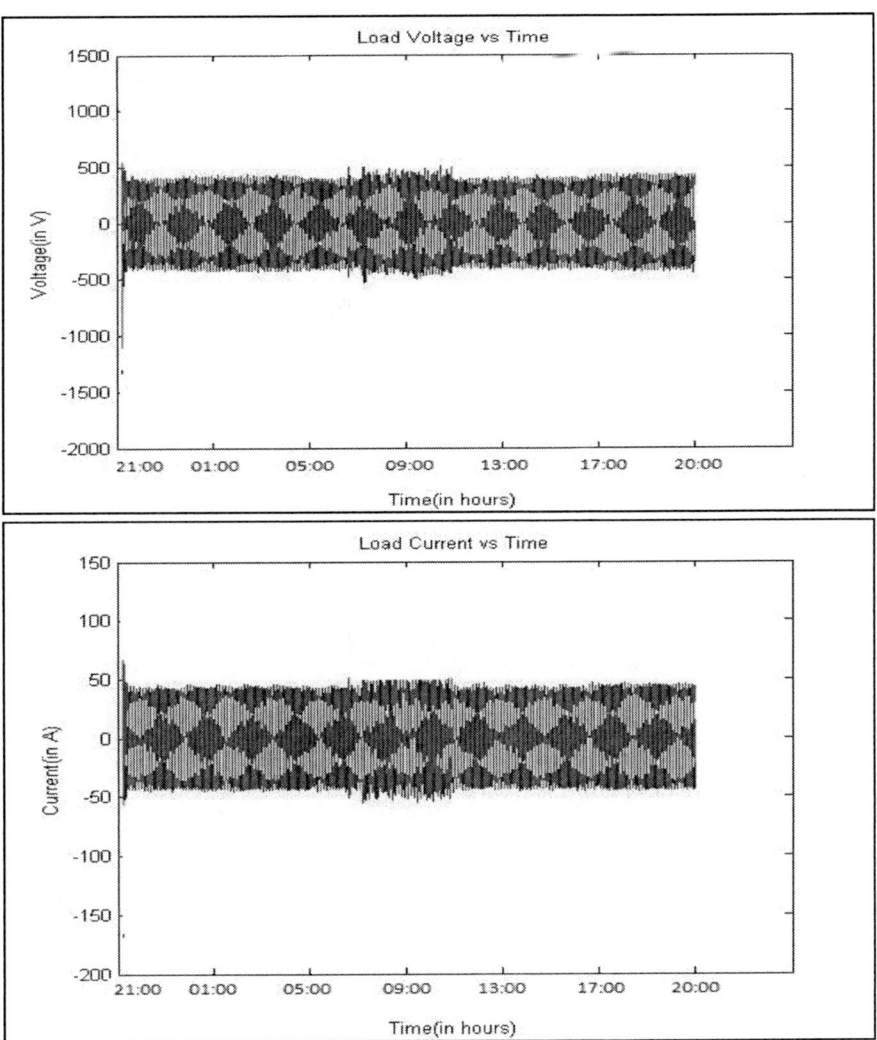

Figure 35. (A) Load voltage (B) load current.

Figure 36 depict the relationship between the irradiance and power for the original data and the optimized data after applying the ABC algorithm. The data was considered for the 24 hours' duration of a sunny day. For optimization the initial data have the parameters like irradiance level, deflection angle, wavelength for a solar array. K-means clustering is used as a fitness function to optimize the irradiance level. It can be clearly seen that optimized irradiance generates slightly more power than the actual irradiance with some minor modifications in the physical system. When the optimized irradiance is used as an input for a parabolic reflector, a much higher power can be obtained thus high load can run. Apart from ABC algorithm many other heuristics algorithm can be used which gives the similar results. Thus, optimization will not only improve the overall efficiency but within the same cost, a good output is also achievable.

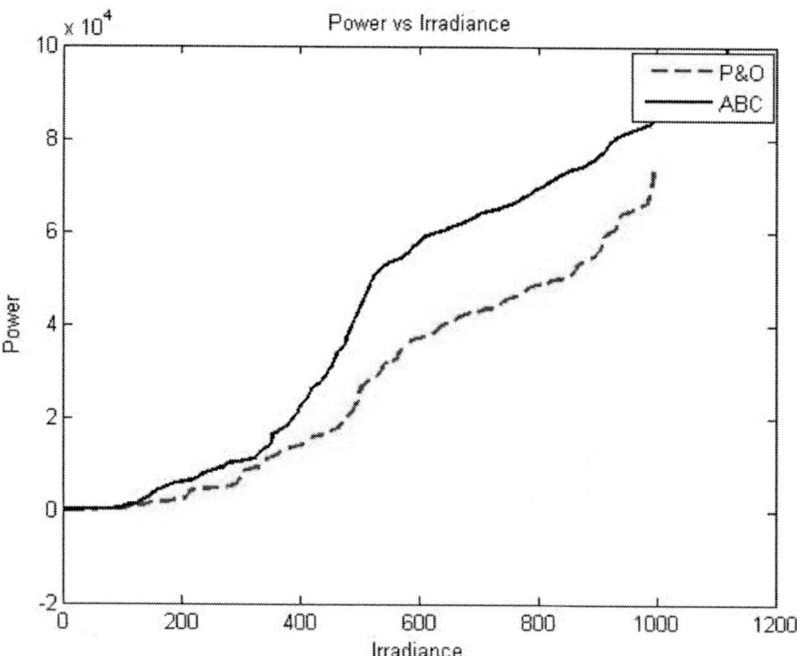

Figure 36. Power versus irradiation graph comparison graph (at instant).

From the graphs it can be seen that with minimized cost and using parabolic reflector and Weibull probability function the overall efficiency can be improved, and more solar irradiance and wind speed can be captured to produce more energy. Table shows the optimized cost using ABC algorithm are placed in APPENDIX 1.

The result shown for specified parameters, which can vary for individual customers, as well as, from area to area.

CONCLUSION

Resulted hybrid energy system appears capable for the location, where grid connection become difficult and commercial fuels are costly due to their cost or untransportable. It is well known fact that the application of hybrid energy system is very useful for rural based area,

Planning of rural electrification area is not an easy task. It involves Technical Economic, Socio –Economic challenges. To evaluate the techno-economics configuration of the system it is very important to develop a proper model and choose the proper optimization algorithm best suited to community. It is very important to get knowledge about load demand and resources in the site.

This chapter proposed the long-term policies of MNRE to electrify the villages of Uttarakhand, India. For optimization of designing the proposed configuration; a modeling and simulation study was carried out in this chapter. Following conclusions have been drawn from this approach:

- Established ABC algorithm works on behavior of honey bee food search approach.
- ABC algorithm has shown better power tracking capability than ABC algorithm and conventional P&O algorithm.
- Voltage and current at PCC has low THD of about 4.5% to 11.95% and is almost uniform and sinusoidal, and able to serve electrical base load.

- Results obtained for P&O algorithm has been found to revel competitive power output for irradiance less than 650 W/m^2 but for irradiance more than 650 W/m^2, proposed algorithm have exhibited better power output.

The established controller is based on active-reactive power flow. Shunt connected capacitor bank provides necessary reactive power demand to regulate voltage. VSC connected with BSU provides active power demand to regulate frequency of system. Voltage regulation has been achieved by adjusting reactive power provided by static compensator consisting of inductor, a VSC and dc bus capacitor. Following conclusions are drawn for this configuration:

- Simulation results shows that voltage and frequency of SEIG-WECS have negligible variation for resistive, reactive, balanced, unbalanced and nonlinear load under varying wind speed and consumer load.
- It eliminates harmonic contents and balances the connected electrical load. So it works as harmonics eliminator and load balancer.
- It is a good power solution for small hydro heads, micro wind turbines and micro grid applications.
- Increase in load power factor increases air gap voltage and speed of prime mover decreases, thereby reducing the frequency of system. It is due to decrease in reactive power requirement with increase in load power factor.
- Established controller regulates voltage and frequency of the system by regulating reactive and active power respectively.
- A good substitute for generating power for low speed variable drive application has been obtained.

In order to plan site for renewable energy-based generation, adequate knowledge of meteorological data and load forecasting information is necessary to carry out techno-economic feasibility study. Data required to interpret optimal utilization of solar and wind resources includes solar irradiance data, solar altitude, solar azimuth, solar angle of incidence, and wind speed at defined altitude. Depending on these factors, installation of generators and requisite components has been decided to ensure optimal results in terms of size, cost, installation area requirement, and energy serving capability. For this purpose, in chapter 4, solar and wind energy and biomass potential in terms of meteorological data and relevant methods to calculate appropriate data has been discussed. Meteorological data of solar energy has been obtained from solar park installed in university campus and meteorological data of wind energy has been taken from Synergy Enviro Engineers (India) Private Limited web data. Following conclusions are drawn towards site meteorological conditions:

- Selected site lies in low wind region and low power wind turbines can be installed for considerable power outcome.
- Considered site is rich in solar resources and can contribute towards significant power contribution.
- Adequate information about solar radiation and temperature is required for noteworthy solar meteorological data for planning installation components.
- In absence of prior knowledge of solar meteorological data, it can be calculated by relevant equations for preceding sunshine hours, radiation data, solar angles and temperature data.
- Performance of PV cell depends on short-circuit current, open-circuit voltage; fill factor, conversion efficiency, clearness index and module maximum power tracking capability.
- In absence of prior knowledge of wind meteorological data, it can be calculated by adequate knowledge about atmospheric conditions like air pressure, air density, wind velocity, wind pattern, temperature and pressure.
- Wind variation and concerned time series data can be generated synthetically by considering methods like Weibull distribution and Rayleigh distribution; and considering factors like Weibull factor 'k' (breadth of wind speeds distribution over the year), 1 hour

autocorrelation factor (wind speed strength in one time step as compared to wind speed in the previous time step), diurnal pattern strength (wind speed strength as compared to time of day), hour of peak wind speed (hour of day that tends to be windiest on average).
- Performance of wind turbine depends on power curve of wind turbine, height of turbine hub and wind pattern over the year.
- Solar irradiation also used as an input for biomass system in MATLAB SIMULATION system.
- All system is working simultaneously. If one source is not available system switches to another.

In order to perform techno-economic feasibility of hybrid PV-wind-bio generation system, real time load data has been investigated and analyzed for Mussoorie area (UTTRAKHAND), Proposed hybrid system lies in low wind region and wind energy has been seen to contribute 8.5% and 76.11% of total generated power. It is found that Hybrid energy system have, such wider circulating than just as individual standalone system for rural remote electrification. However, there are several kinds of hybrid energy system, the most suitable of which for the application in small community depends on the climatic conditions at the application field. It is also shown that combining more than one energy supply system is a better solution for the remote application.

Several drawbacks limit the progress of hybrid energy system. The politics of energy deployment, familiarity of this technology and education are some salient issues. Due to this, the role which renewable energy-based hybrid energy systems play in meeting the increased demand for clean electricity and assisting economic development is not fully appreciated and has largely been ignored in national plans. Hence, there is a lack of confidence in this technology.

With good resources and demand estimation, system sizing, economic analysis, operation and maintenance practices, hybrid energy system in Indian national rural remote region are feasible, viable option with the added benefit of being environmentally friendly.

Considerable amount of wind power has been generated during the month of April to September.

- Site location is rich in solar resource and PV power has contributed to about 64.0% wind system contributed 35.0% and bio mass contributed only 1.0% total generated power.
- No fossil fuel-based power source has been considered in this study and thus renewable fraction is .07% but when no fossil fuel used renewable fraction is 100%.
- The COE found in optimization result is Rs.45.46 and Rs.19.45in case study 1 (with fossil fuel) and case study 2 (without fossil fuel) respectively, which is less than present COE (grid connection and diesel generator) of Rs. 20.20and likely to increase in near future as per decreasing COE trend of RES based generation.
- Obtained COE is competitive COE with hybrid PV-wind-bio-diesel system Rs.19.45/kWh,hybrid PV-Wind-Diesel Rs.13.46/ kWh,hybrid PV-bio-diesel Rs. 17.71/kWh, hybrid wind- diesel Rs.14.21/kWh and hybrid wind –bio-diesel Rs. 33.23 /kWh.
- The optimized penetration level for wind energy at 32.6 m hub height and 12.5 scale factors with 1.4 surface friction and rotor diameter 19.96 m^2 at 310499712.8 Rs cost.
- Biomass system is working on the solar energy input energy for biomass system is taken from solar system.
- It is concluded that the proposed hybrid system can be supply the villages' population with 24 h power supply @ Rs. Rs.8-10/kWh form solar system, Rs 9/kWh from wind system and around 6/kWh form biomass system.
- As such we are using only the resistive load, we don't have any of the capacitive or inductive load. So the quality of power will be active

- Rate of energy generation from photovoltaic system is 66%
- Rate of energy generation from wind system is 56%
- Rate of energy generation from biomass system is 28%
- Power factor not dropped below 0.9

Solar energy needs larger space and needs to be established in open space to get exposure of sun light. These constrains does not affect wind turbines having sufficient ground clearance. Despite of lying in low wind region, proposed site is suitable for potential generation from RES. COE of renewable energy-based generation approaches have been found to follow reducing trend in last decade due to improved efficiency and technical upgradation of conversion components. On the other hand, fossil fuel-based generation have revealed increasing COE trend due to depleting reserves. Lifetime of the components has been taken to standard rated period as prescribed by manufacturers and components can be used further with little or no maintenance, thus reducing net COE further. Lifetime of the components has been taken to standard rated period as prescribed by manufacturers and components can be used further with little or no maintenance, thus reducing net COE further.

APPENDIX

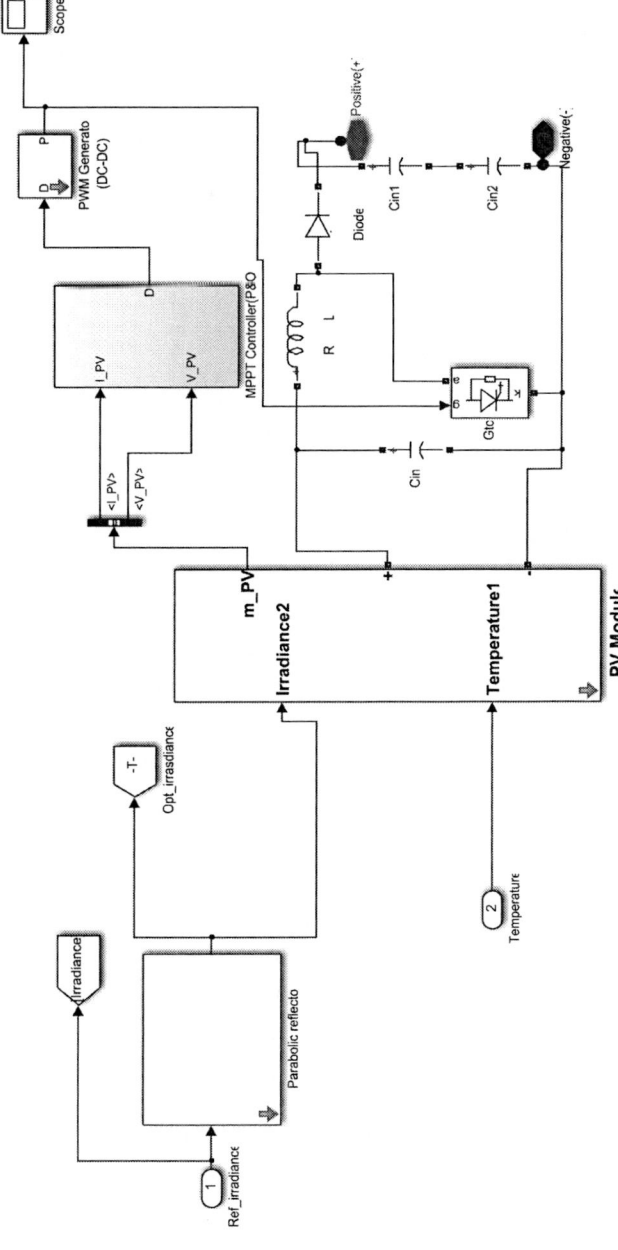

Figure A1. Simulation model of PV.

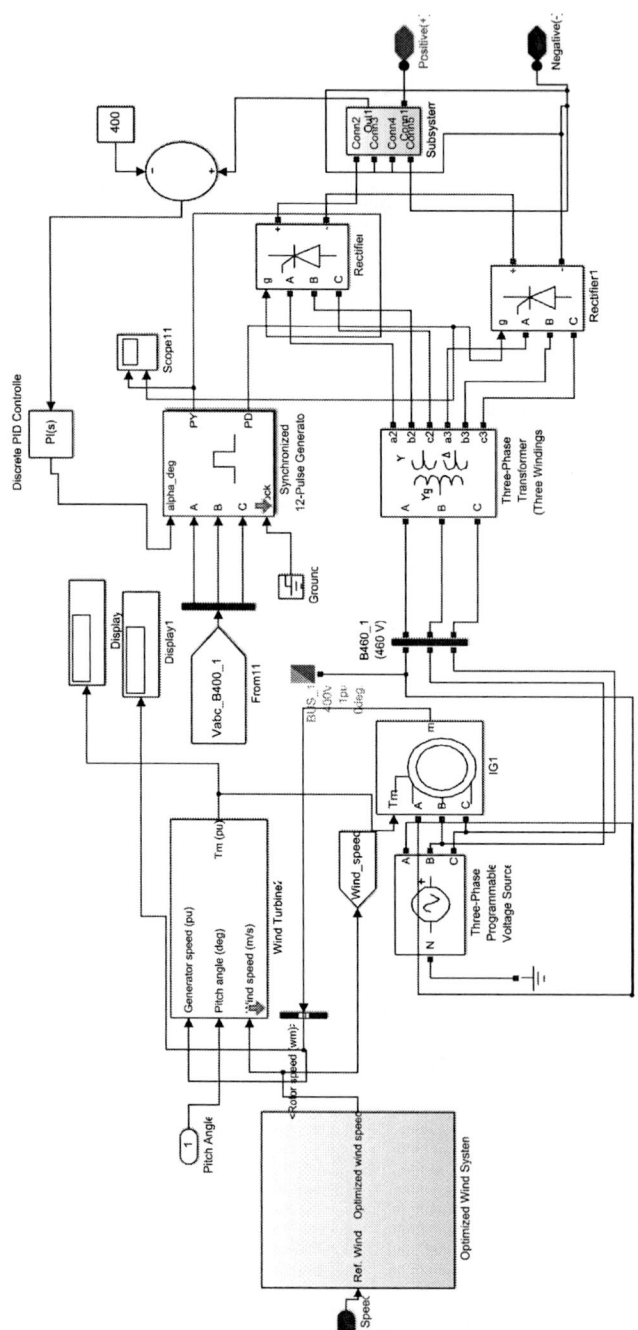

Figure A2. Simulation of WT.

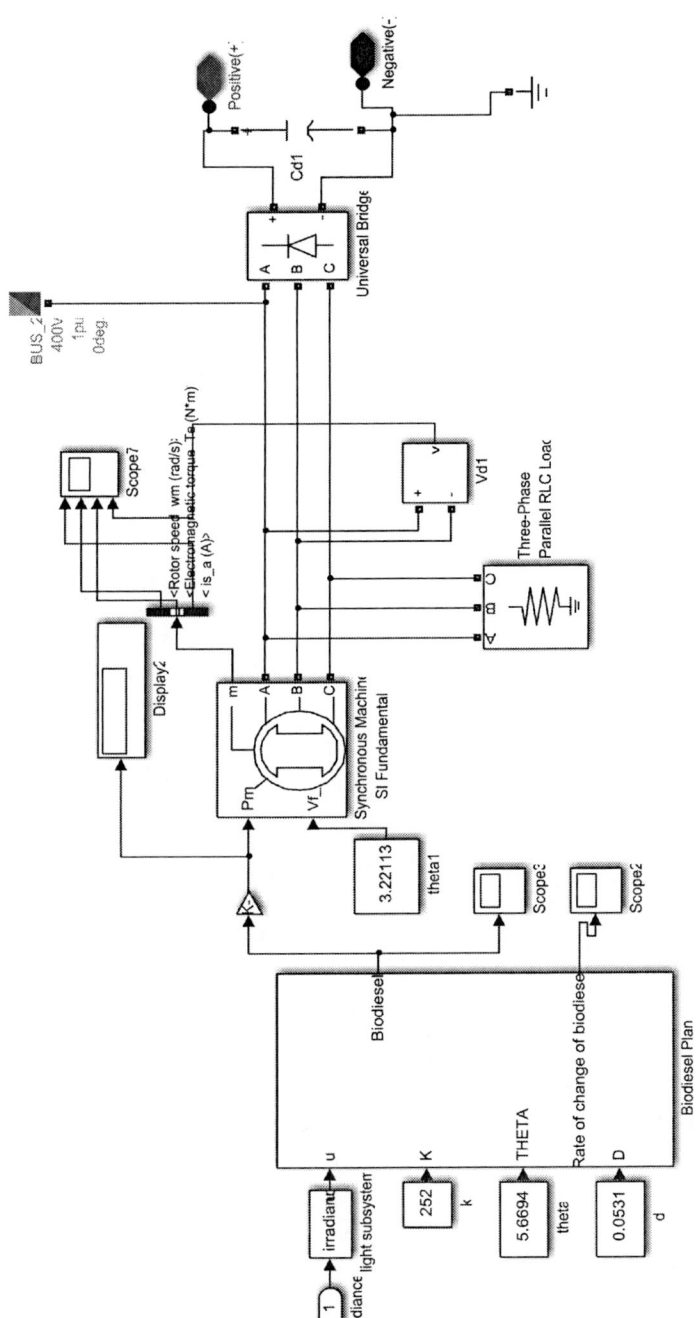

Figure A3. Simulink model of biomass.

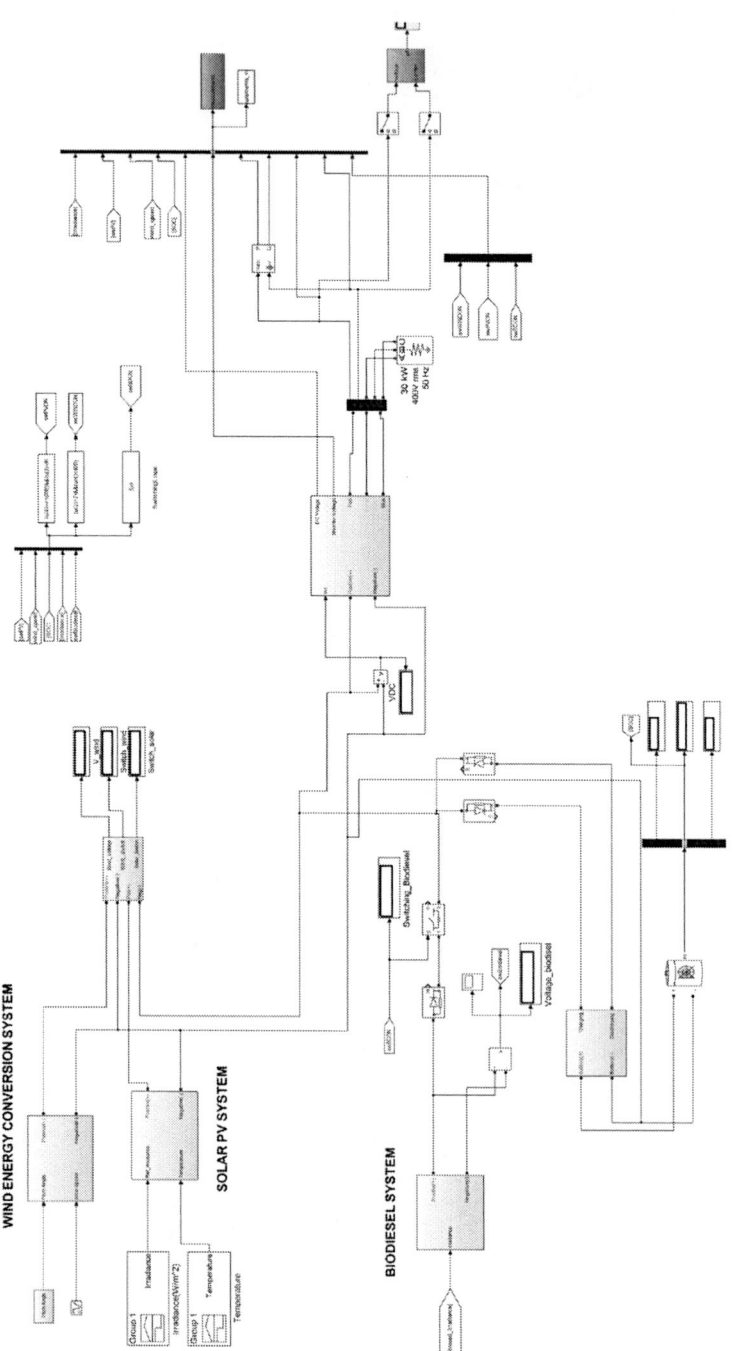

Figure A4. Complete Simulink model for HRES.

REFERENCES

[1] E. I. Zoulias, and N. Lymberopoulos. 2007. "Techno-economic analysis of the integration of hydrogen energy technologies in renewable energy-based stand-alone power systems." *Renewable Energy, 32*(4): 680-696.

[2] A. Gupta, R. P. Saini, M. P. Sharma. 2011. "Modelling of hybrid energy system—Part I: Problem formulation and model development." *Renewable Energy 36*(2):459-465.

[3] A. Gupta, R. P. Saini, M. P. Sharma. 2011. "Modelling of hybrid energy system—Part II: Combined dispatch strategies and solution algorithm." *Renewable Energy 36*(2):466-473.

[4] A. Gupta, R. P. Saini, M. P. Sharma. 2008. "Computerized Modeling of Hybrid Energy Systems Part III: Case Study with Simulation Result." *IEEE conference on Electrical Engineering* 7-12.

[5] A. Gupta, R. P. Saini, M. P. Sharma. 2008. "Hybrid energy system for remote area-an action plan for cost effective power generation." In *Industrial and Information Systems, 2008. ICIIS 2008. IEEE Region 10 and the Third international Conference* 1-6.

[6] H. Fathima, and K. Palanisamy. 2015. "Optimization in microgrids with hybrid energy systems–A Review." *Renewable and Sustainable Energy Reviews 45*: 431-446.

[7] S. Chin, A. Babu, W. McBride. 2011. "Design, modeling and testing of a standalone single axis active solar tracker using MATLAB/Simulink." *Renewable Energy 36*(11): 3075-3090.

[8] O. Hafez, K. Bhattacharya. 2012. "Optimal planning and design of a renewable energy based supply system for microgrid." *Renewable Energy 45*:7-15.

[9] A. H. Fakehi, S. Ahmadi, M. R. Mirghaed. 2015. "Optimization of operating parameters in a hybrid wind–hydrogen system using energy and exergy analysis: Modeling and case study." *Energy conversion and management 106*:1318-1326.

[10] K. H. Chang, and G. Lin. 2015. "Optimal design of hybrid renewable energy systems using simulation optimization." *Simulation Modelling Practice and Theory 52*:40-51.

[11] Tsuanyo, Y. Azoumah, D. Aussel, P. Neveu, P. 2015. "Modeling and optimization of batteryless hybrid PV (photovoltaic)/Diesel systems for off-grid applications." *Energy 86*:152-163.

[12] W. K. Yap, and V. Karri. 2015. "An off-grid hybrid PV/diesel model as a planning and design tool, incorporating dynamic and ANN modelling techniques." *Renewable Energy 78*:42-50.

[13] T. Ahmad. 2016. "A hybrid grid connected PV battery energy storage system with power quality improvement." *Solar Energy 125:* 180-191.

[14] Y. A. Katsigiannis, P. S. Georgilakis, E. S. Karapidakis. 2012. "Hybrid simulated annealing–tabu search method for optimal sizing of autonomous power systems with renewable." *IEEE Transactions on Sustainable Energy 3*(3):330-338.

[15] M. Bashir, and J. Sadeh. 2012. "Optimal sizing of hybrid wind/photovoltaic/battery considering the uncertainty of wind and photovoltaic power using Monte Carlo," *Environment and Electrical Engineering (EEEIC) 2012 11th International Conference* 1081-1086.

[16] T. Khatib, A. Mohamed, K. Sopian. 2012. "Optimization of a PV/wind micro-grid for rural housing electrification using a hybrid iterative/genetic algorithm: case study of Kuala Terengganu, Malaysia." *Energy and Buildings* 47:321-331.

[17] Abbes, A. Martinez, G. Champenois. 2014. "Life cycle cost, embodied energy and loss of power supply probability for the optimal design of hybrid power systems." *Mathematics and Computers in Simulation* 98:46-62.

[18] R. K. Rajkumar, V. K. Ramachandaramurthy, B. L. Yong, D. B. and Chia. 2011. "Techno-economical optimization of hybrid pv/wind/battery system using Neuro-Fuzzy." *Energy* 36(8):5148-5153.

[19] J. M. Lujano-Rojas, R. Dufo-López, J. L. Bernal-Agustín.2 013. "Probabilistic modelling and analysis of stand-alone hybrid power systems." *Energy* 63:19-27.

[20] M. Sanchez, A. U. Chavez-Ramirez, S. M. Duron-Torres, J. Hernandez, L. G. Arriaga, J. M. Ramirez. 2014. "Techno-economical optimization based on swarm intelligence algorithm for a stand-alone wind-photovoltaic-hydrogen power system at south-east region of Mexico." *International journal of hydrogen energy* 39(29):16646-16655.

[21] A. Maleki, F. Pourfayaz. 2015. "Optimal sizing of autonomous hybrid photovoltaic/wind/battery power system with LPSP technology by using evolutionary algorithms" *Solar Energy 115*: 471-483.

[22] K. H. Chang, K. H., and G. Lin, G. 2015. "Optimal design of hybrid renewable energy systems using simulation optimization." *Simulation Modelling Practice and Theory* 52: 40-51.

[23] E. Türkay, and A. Y. Telli. 2011. "Economic analysis of standalone and grid connected hybrid energy systems." *Renewable energy* 36(7):1931-1943.

[24] Prakash Kumar, Gaurav Jain, and Dheeraj Kumar Palwalia. 2015. "Genetic algorithm based maximum power tracking in solar power generation." In *Power and Advanced Control Engineering (ICPACE), 2015 International Conference on* 1-6.

[25] G. Tzamalis, E. I. Zoulias, E. Stamatakis, E. Varkaraki, E. Lois, F. Zannikos. 2011. "Techno-economic analysis of an autonomous power system integrating hydrogen technology as energy storage medium." *Renewable energy* 36(1):118-124.

[26] A. Ghasemi, A. Asrari, M. Zarif, S. Abdelwahed. 2013. "Techno-economic analysis of stand-alone hybrid photovoltaic–diesel–battery systems for rural electrification in eastern part of Iran—A step toward sustainable rural development." *Renewable and Sustainable Energy Reviews* 28:456-462.

[27] M. A. Ramli, A. Hiendro, S. Twaha. 2015. "Economic analysis of PV/diesel hybrid system with flywheel energy storage." *Renewable Energy* 78:398-405.

[28] W. Zhou, C. Lou, Z. Li, L. Lu, H. Yang. 2010. "Current status of research on optimum sizing of stand-alone hybrid solar–wind power generation systems." *Applied Energy* 87(2):380-389.

[29] O. Erdinc, and M. Uzunoglu. 2012. "Optimum design of hybrid renewable energy systems: Overview of different approaches." *Renewable and Sustainable Energy Reviews* 16(3):1412-1425.

[30] H. Belmili, M. F. Almi, B. Bendib, S. Bolouma. 2013. "A Computer program development for sizing stand-alone photovoltaic-wind hybrid systems." *Energy Procedia* 36:546-557.

[31] M. Castaneda, A. Cano, F. Jurado, H. Sánchez, L. M. Fernandez. 2013. "Sizing optimization, dynamic modeling and energy management strategies of a stand-alone PV/hydrogen/battery-based hybrid system." *International journal of hydrogen energy* 38(10): 3830-3845.

[32] H. Yang, Z. Wei, L. Chengzhi. 2009. "Optimal design and techno-economic analysis of a hybrid solar–wind power generation system." *Applied Energy*, 86(2):163-169.

[33] O. Ekren, and B. Y. Ekren. 2010. "Size optimization of a PV/wind hybrid energy conversion system with battery storage using simulated annealing." *Applied Energy* 87(2):592-598.

[34] M. J. Khan, and M. T. Iqbal. 2005. "Dynamic modeling and simulation of a small wind–fuel cell hybrid energy system." *Renewable energy* 30(3): 421-439.

[35] O. C. Onar, M. Uzunoglu, M. S. Alam. 2006. "Dynamic modeling, design and simulation of a wind/fuel cell/ultra-capacitor-based hybrid power generation system." *Journal of power sources*, 161(1): 707-722.

[36] M. Veerachary. 2005. "Power tracking for nonlinear PV sources with coupled inductor SEPIC converter." *IEEE transactions on aerospace and electronic systems* 41(3):1019-1029.

[37] N. Pandiarajan, and R. Muthu. 2011. " Mathematical modeling of photovoltaic module with Simulink." *International Conference on Electrical Energy Systems* 6.

[38] R. Dufo-Lopez, J. L. Bernal-Agustín, J. Contreras. 2007. "Optimization of control strategies for stand-alone renewable energy systems with hydrogen storage." *Renewable energy* 32(7):1102-1126.

[39] www.mnre.gov.in (Accessed on 2/January2018.

[40] www.cea.nic.in (Accessed on 2/January2018).

[41] www.makeinindia.com (Accessed on 2/January2018).

[42] www.cercind.gov.in (Accessed on 2/January2018).
[43] www.ireda.gov.in (Accessed on 2/January2018).
[44] Renewable energy "Akshay Urja", newsletter, ministry of new and renewable energy, vol. 6 issue 4 February 2013.
[45] Report No. 44/26/05-RE (Vol-II) (2006),"rural electrification policy," ministry of power, India.
[46] https://data.gov.in/visualize3/?inst=f859972622261c0f3637ce358ef22ba5andvid=9101# accessed on 31 may 2016.
[47] http://www.ddugjy.gov.in/mis/portal/dcovered.jsp?stcd=05 accessed on 31 may 2016.

BIOGRAPHICAL SKETCH

Shweta Goyal

Affiliation: Graphic Era Deemed to Be University

Education: PhD scholar (submitted)

Business Address: Graphic Era Deemed to Be University

Research and Professional Experience: PhD scholar

Publications in Last Three Years
1) Shweta Goyal, Sachin Mishra, Anamika Bhatia "Optimization of Hybrid Energy System" in *International Conference on Intelligent Circuits and Systems (ICICS)* (pp. 349-354). IEEE. April 2018
2) Shweta Goyal, Sachin Mishra, Anamika Bhatia "Optimization of size of PV/wind/biodiesel by using Artificial Bee Colony (ABC) algorithm" in In *Control, Automation & Power Engineering (RDCAPE), 2017 Recent Developments in* (pp. 220-223). IEEE October 2017

3) Shweta Goyal, Sachin Mishra, Anamika Bhatia "A comparative approach between different optimize result in hybrid energy system using HOMER" in *International Journal of Electrical and Computer Engineering (IJECE); 9[1];141-147; February 2019.*
4) Shweta Goyal, Sachin Mishra, Anamika Bhatia "Electricity Generation with Conventional and Non-Conventional Energy Technologies in Dudhali Village" in *"Indian Journal of Science and Technology."* August 2017; 10[30].
5) Shweta Goyal, Sachin Mishra, Anamika Bhatia "Modeling of solar/wind hybrid energy system using MATLAB Simulink" in *"International Journal of Pharmaceutical Research / Special Issue."* 9[3]; September 2017.
6) Shweta Goyal, Sachin Mishra, Anamika Bhatia "Optimization Strategies for Hybrid Energy System: A Review, in *International Journal of Applied Engineering Research;* 13(9); 7010-7018; 2018.
7) Shweta Goyal, Sachin Mishra, Anamika Bhatia "A Review on Energy Policies and Scenario in India" in *"International Journal of Advances in Applied Sciences,* 6(2), 156-161 June 2017.

International Conference
1) Shweta Goyal, Sachin Mishra, Anamika Bhatia "Electricity Generation with Conventional and Non-Conventional Energy Technologies in Dudhali Village" in *"RED SET 2016" Springer conference.*
2) Shweta Goyal, Sachin Mishra, Anamika Bhatia "Optimization of size of PV/wind/bio mass by using Artificial Bee Colony (ABC) algorithm *in RDCAPE 2017 IEEE conference*
3) Shweta Goyal, Sachin Mishra, Anamika Bhatia, Puspender Singh Jarodia "Hybrid Energy System for A Rural Area in *RIEECS 2017 conference.*
4) Shweta Goyal, Sachin Mishra, Anamika Bhatia "Different component optimization of hybrid energy system" in ICICS2018 *IEEE conference.*

National Conference
1) Shweta Goyal, Sachin Mishra, Anamika Bhatia "Modeling of Solar/Wind Hybrid Energy System Using MATLAB Simulink" in *NCRTGCS 2017* held at Graphic Era University.
2) Shweta Goyal "simplified approach to check the stability time and frequency domain using MATLAB" in *NCRAEEE 2012* held at Dev Bhoomi institute of Technology.

Book Chapter (Accepted)
1) Shweta Goyal, Sachin Mishra, Anamika Bhatia "Techno Economic Analysis of Solar-Biomass-Diesel-Battery-Hybrid Energy System for a Remote Area ." is accepted by Editor for book Advanced Engineering Research and Applications (AERA) "Chapter Code: 1211"

Poster
1) Shweta Goyal, Sachin Mishra, Anamika Bhatia "Different Component Optimization of Hybrid Energy System" in 12th Uttarakhand State Science and Technology Congress 2017-18.

In: A Comprehensive Guide to Energy ... ISBN: 978-1-53616-728-3
Editor: Shannon Alvarado © 2020 Nova Science Publishers, Inc.

Chapter 5

DESIGN AND MANUFACTURING APPROACHES TO IMPROVE RELIABILITY OF WIND TURBINE GEARBOXES: A REVIEW

Wei Jiang[*]
School of Mechanical Engineering,
Dalian University of Technology,
Dalian, P. R. China

ABSTRACT

Renewable energy has gained increasing interest worldwide against energy crisis and environmental pollution. As a renewable and clean source of energy, wind energy power generation has had an exponential growth in the last decade. However, for a typical design life of 20 years of wind turbines, the premature failure of wind turbine gearboxes greatly increase the downtime, maintenance expenditures, and cost of energy.

Wind turbine gearboxes are one of the most expensive components in wind turbine system. Operational experiences reveal that gearboxes are the weakest link in multi-megawatt wind turbines. This paper first studies the

---

[*] Corresponding Author's E-mail: jiangwei@dlut.edu.cn.

failure modes of wind turbine gearboxes and their causes. Factors affecting the gearbox reliability in both design and manufacturing processes are then analyzed. Measures to improve the reliability and prolong the service life of MW sized wind turbine gearboxes through innovative design and manufacturing approaches are finally proposed. Hopefully, by advanced design and manufacturing, high reliability gearboxes can be developed to operate confidently under complex loading conditions.

Keywords: wind turbine gearbox, failure modes, reliability, design and manufacturing, review

1. INTRODUCTION

Renewable energy has gained increasing interest worldwide against energy crisis and environmental pollution. As a renewable and clean source of energy, wind energy power generation has had an exponential growth in the last decade (Igba et al. 2017). At the same time, the rated power capacity of wind turbines has increased greatly, from 15 kW in the 1980s to around 9.5 MW in 2015 (Jantara et al. 2018). The expansion of wind power capacity not only increases the size of wind turbines but also increases the likelihood of failure. For a typical design life of 20 years of wind turbines, the premature failures of wind turbine components decrease the operational time and increase the downtime and maintenance expenditures.

Wind turbines convert mechanical energy to electrical power. The main drive chain in a wind turbine consists of blades, low-speed input shaft, bearings, gears, high-speed output shafts, and a generator. The blades are connected to the rotor by hub. Driven by the wind, the blades transmit the torque via the low speed shaft through the gearbox to the high speed shaft that is attached to the generator.

Gearboxes convert the rotational speed of rotor blades from 5 to 22 rpm to the generator-required rotational speed of around 1000 to 1600 rpm (Ragheb and Ragheb 2010). Considering the limitation of space and the requirement of efficiency, approximately 75% of commercial wind turbines use three-stage gearboxes (Jantara et al. 2018), employing a planetary gearing system. The gearbox increases the input rotational speed to the speed required by the generator while reducing the transmitted torque by the same ratio. As a main transmission component, the performance of gearbox relates closely to the safety of wind turbine operation.

Unlike conventional gearboxes used in steady-state applications, wind turbine gearboxes usually operate in harsh, adverse environments, such as in deserts, offshore and mountains. In addition to facing the long operational time, wind turbine gearboxes confront high failure rate. Operational experiences reveal that gearboxes are the weakest link in the multi-megawatt wind turbines. (Ragheb and Ragheb 2010). Few of them could reach an operational period of more than 7 years.

Gearboxes are important components connecting the hub and generator. The failure of gearbox will cause significant downtime, malfunction of wind turbine, and reduction of power generation. Since an average size wind turbine gearbox is cost prohibitive (~$500 K) (Willis et al. 2018), the repair or replacement of a gearbox can result in significant expenditure, cost up to 10% of the original construction cost (Ragheb and Ragheb 2010), greatly cut the projected profits and increase the cost of wind energy.

Failure refers to the incapability of a machine or an element to perform its intended function (Mott 2003). The failure of wind turbine gearboxes are caused by many factors, and most of them can be traced back to the design and manufacturing quality. However, statics analyses show that the same type and power wind turbines present different failure rate when they operate at different location and/or weather conditions, implying that operation conditions, such as extreme weather, long-term alternating loads, and their combinations, are also responsible for the failure of wind turbine gearboxes.

Researchers worldwide have been investigating the problem of wind turbine gearboxes over the past 20 years (Igba et al. 2015). National Renewable Energy Lab (NREL) engaged researchers, bearing, gearbox and wind turbine manufacturers, consultants, and owners of wind turbines to initiate Gearbox Reliability Collaborative (GRC) research to address gearbox reliability in 2007 (Musial, Butterfield, and McNiff 2007). Organizations and universities worldwide, such as Sandia National Laboratory, RisØ DTU, China Power Group, Delft University of

Technology, Norwegian University of Science and Technology, etc., have also made great progresses through extensive and continuous researches.

Once the faults of gearbox can be identified in time and handled properly, the operation life of the gearbox could be prolonged and the safety can be guaranteed. Therefore, it is of great significance to study the main root causes of failure and propose recommendations for the design, manufacturing and operation of wind turbine gearboxes.

This paper first introduces the failure modes in wind turbine gearboxes. Factors affecting the failure rate are then analyzed and the root cause of failure are identified for a better understanding of the fundamental failure mechanisms. The options and measures to improve the reliability and prolong the service lifetime of MW sized wind turbine gearboxes through design, manufacturing and maintenance approaches are finally proposed.

2. Failure Modes of Wind Turbine Gearboxes and Their Causes

Failures in wind turbine gearboxes have been studied widely in order to collect reliability data to improve design and manufacturing. Among various terms describing failures, failure rates as failures per turbine per year, and downtime as hours lost per component per turbine per year, are the most commonly used terms. According to the statistics of quality problems investigated, failures in a wind turbine gearbox occur on bearings, gears, planetary frame, housing, and lubrication and cooling system. This section sums up failures in the main components, and analyze the root cause of failures. Suitable solutions to these problems could reduce failure rate and help identify the best design and manufacturing approaches to increase service reliability.

2.1. Bearing Failures

Bearings are found to be the leading source of failure in a wind turbine gearbox (Musial, Butterfield, and McNiff 2007). They are identified as the root cause of the majority of component failures in gearboxes. Initiated at several specific locations, bearing debris may subsequently progress into gear meshing teeth, producing surface wear and causing misalignments of gears (Perez et al. 2013; Lin et al. 2016).

Bearing failure modes include fatigue pitting, abrasion and adhesion wear, cracking, corrosion, overload, overheating, and other phenomena (Greco et al. 2013). These failures occur due to improper installation, insufficient lubrication, lubricant pollution and adverse working environments.

Bearings in wind turbine gearboxes operate under adverse loading conditions, requiring adequate sealing and lubrication to ensure long service life. In low-speed bearings, such as spindle bearings whose rotational speed is between 10-20 rpm, stable and reliable lubrication film is hard to form. Therefore, under the condition of downwind, the contact surface of roller and raceway slides under heavy load and thin oil film, leading wear on bearing surfaces.

In high-speed bearings, micro cracks may initiate under the cyclic Hertzian contact stress. The micro cracks propagate, causing fatigue pitting, spalling and surface wear. The wear debris may further affect bearing clearance and the mating of bearing.

When onshore wind turbines operate in dusty environment, sand, rust, together with metal chips, and wear debris are common contaminations. The hard, sharp-edges particles in the lubricant, as well as asperities on one bearing surface scratch the mating bearing surface, especially the low-speed shaft surface, where good lubrication is difficult to establish due to low-speed and heavy-load condition. Insufficient lubrication may raise temperature and aggravate wear damage. Poor lubrication also cause severe adhesion or scuffing, where material from one bearing surface is transferred to another due to local micro welding and tearing.

Offshore wind turbines operate in the corrosive environment. Apart from the above mentioned failure modes, bearings may fail due to corrosion.

2.2. Gear Failures

Defects induced in manufacturing process may cause gear failures; however, most of gear failures originate from fatigue. Like conventional gearboxes, gears in wind turbine gearboxes may fail due to gear tooth fracture and tooth surface damage. Tooth fracture includes overload fracture and low/high cycle bending fatigue fracture; while tooth surface damage includes Hertzian pitting fatigue, abrasion, adhesion, scuffing, gluing, etc. (Greco et al. 2013).

Teeth fracture may be caused by sudden overload or gradual propagation of micro cracks due to cyclic loading. The load on wind turbine gearboxes varies greatly. The overload caused by extreme wind speeds, or by the instantaneous peak load due to emergency braking, maybe several times greater than the rated load, resulting in overload fracture of gear teeth. In addition, bearing damage, shaft bending, excessive local stresses, and stress concentration may also cause gear teeth fracture.

High cycle bending fatigue fracture occurs when the gear teeth experience cyclic stresses. The planetary gear is more likely to occur fatigue pitting and tooth breakage as it takes bidirectional forces during operation. The small sun has low accuracy and carries large loads, and is easy to induce failure. Fatigue bending usually initiates near root fillets, where stresses are high or where stress concentration occurs. Micro cracks propagate under cyclic loads, resulting in either ductile, brittle or mixed-mode fracture depending on material toughness and magnitude of stresses.

Tooth surface damage is the most frequently occurred failure. The initial micro cracks may propagate under cyclic Hertzian contact stresses, resulting in pitting and gradually expand to spalling due to excessive cyclic contact stresses under alternating loads.

Impurities entering the gearbox during manufacturing, assembly or maintenance, debris produced by wear, or foreign hard particles admitted by seals, may lead to abrasive wear on the meshing gear teeth surfaces. When lubricant is inadequate, the contact surfaces of a pair of meshing gears are pressed together under cyclic Hertzian contact stresses, causing metal-to-metal contact and adhesion of surface asperities, resulting in adhesion, scuffing, gluing (Musial, Butterfield, and McNiff 2007). The use of surface-hardened gears and high viscosity lubricants can minimize the generation of wear debris.

At early stage of operation, when a small amount of pitting or slight wear occur, the damage of gear tooth surface can be alleviated by replacing the lubricant or by adjusting lubrication mode. When cracks, broken teeth, gluing, or considerable wear occur, the gear should be repaired or replaced.

2.3. Lubrication Failures

Failures relate to lubrication include insufficient lubrication, oil leakage, the faults in oil pump, heater, cooler, filter, etc. Oil starvation is one of direct root causes of the failure of most gearbox components. Poor lubrication may result in friction, wear, and increase of temperature, leading to premature failure of gears and bearings. Wear debris may further contaminate oil and reduce the service life of gear lubricants (Nikas 2010).

Wind turbines working in either cold or hot climates require a heating and cooling system be properly designed to keep lubricant oil at constant temperature. Temperature closely relates to lubricant viscosity, i.e., too high a temperature may result in low viscosity and thin oil film thickness, easy to cause friction and wear; while too low a temperature may result in high viscosity and cause overload or malfunction of oil pumps.

Oil leakage in a gearbox can be attributed to the structural defects under adverse working conditions. Low temperature or vibration loads are likely to cause oil leakage at the interface between housing and gear ring, end cover and housing, journals of low-speed shaft and high-speed shaft, joints of lubrication and cooling system. The possible causes of oil leakage may be the loosening of bolts, unreasonable structure design, improper selection of sealing parts, etc. At the same time, the failure of seal may easily let external dust enter the gearbox and contaminate lubricant oil.

2.4. Housing Failures

Multi-megawatt wind turbine gearboxes usually have large size and weight. For example, the mainstream 3 MW gearbox usually have the length and width of more than 3 m, and the weight of more than 20t, with the input torque of about 2 MN·m (Wang et al. 2013). At this scale, material processing, manufacturing, transportation, and assembly may easily cause problems.

Wind turbine gearboxes usually carry much more complex loads than conventional gearboxes. They are usually subjected to variable loads, combined with extreme loads and impact loads. The extreme load may lead the gearbox of 1.5-3.0 MW rated power wind turbine subjected to 3-4 times of nominal rated power (Wang et al. 2013), which greatly affects those low stiffness or low strength components. Under complex loads, especially when the working load exceeds the design load, the large torque introduced will magnify the eccentricity error or assembly error, inducing vibration, stress concentration and even structural failure.

The main failure modes of a gearbox housing are fracture, deformation, wear of bearing seats, etc. The housing damage mostly appears at the connection between the housing and the gear ring, where the bolts connecting the housing and the gear ring are seriously broken. Besides, when the axial positioning of the rotating shaft fails, the rotating shaft may cause wear on the bearing seats.

Vibration is another kind of failure mode of a gearbox housing, caused by low accuracy of gearbox or large assembly errors of components. Besides, low strength gearbox housing and insufficient shaft stiffness will lead to the deformation of gear shaft and affect gear meshing.

2.5. Failures of Other Components

The main failure modes of shafts are fracture, deformation, and wear. When a shaft fails, it will affect power transmission, cause damage to other elements, and therefore, the machine must be stopped immediately, and the shaft be repaired or replaced.

The main failure modes of planetary frame are fracture and wear. The damage in the planetary frame often occurs at low strength region, such as planetary hole, or the area close to planetary wheel or planetary frame bearing.

Bolts of gearbox at the joint of gearbox housing and gear ring usually fail due to breakage. The impact loads, poor heat treatment and surface quality, bolt loosening, or improper tightening torque, are responsible for such failure.

Apart from the previously discussed individual failure modes, components in wind turbine gearboxes frequently fail due to multiple failure mode, i.e., two or more failure modes occur simultaneously. Individual failures experienced by a component may interact with each other, or one failure may trigger another. Multiple failure mode makes the analysis of failure mechanism even more complicated.

In summary, as the power is transmitted from the blade to the low-speed input shaft, bearings, gears, high-speed output shaft, and generator, failure may occur in any element individually, or occur in several elements sequentially. Gearboxes, together with the yaw system, pitch system, hydraulic system and lubrication system, form a complex system, which brings great challenges to the failure analyses.

3. DESIGN APPROACHES TO IMPROVE RELIABILITY OF WIND TURBINE GEARBOXES

3.1. Load Prediction

One of the main differences between a wind turbine gearbox and a conventional gearbox is the load they carry. Wind turbines rely on wind to work. During operation, the wind speed varies from 3 to 25m/s. The fluctuating wind loads produce progressive damage that can ultimately result in wind turbine structural failure (Sørensen, Unnikrishnan, and Mathew 2001).

Parameters describing wind climate include the annual mean wind speed, the Weibull distribution of wind speed, extreme wind speed, wind shear, air density, flow inclination, etc. (Toft et al. 2016, Hulio 2019). The precise prediction of wind load involves a large number of events during normal operation, as well as additional cases of extreme and accidental events. Veers and Winterstein (1998) illustrated a statistical approach to analyze and process measured wind parameters to satisfy the needs of fatigue-life prediction, loading-spectra definition, and reliability estimates.

A gearbox, as the core component of a wind turbine, is connected to the blades at one end and the generator at the other end. During operation, the wind gearboxes are subjected to difference kinds of loads. Externally, they are impacted by the aerodynamic fluctuating torque transmitted through the blade, oscillating considerably under the action of gust wind, plus the excitations induced by grid oscillations, and rotor mass imbalance, etc. (Nejad et al. 2016). Internally, they are subjected real-time loads due to misalignments, bending moments, temperature and humidity variations. These external and internal loads form a wide range of possible load cases that comprise the design load spectrum.

Early wind turbine gearboxes were designed according to static loads which under-estimated the effect of variable operating loads on the life of gearboxes. Although considering dynamic loads could produce a robust design, the variable turbine loads is hard to predict accurately. The lack of satisfactory understanding and prediction of the stochastic and cyclic loads are the reasons that prevent the majority of wind turbine gearboxes from reaching their expected service lifetime of 20-25 years.

Currently, external alternating loads on the wind turbine gearboxes can be calculated by commercially available or public domain software. A decoupled analysis method is used to estimate the dynamic load response from the external load (Nejad et al. 2016). Multi-body dynamic analysis is used to model wind turbine loads, coupled with internal loads and deformations of the gearbox to form the total loads on the gearbox components.

Aeroelastic calculation codes calculate design loads and power as a function of time. Multi-body dynamic analysis can be performed to relate the rotor loads calculated by the FAST wind turbine code to the multi-body dynamics of gears and shafts (Wilson, Walker, and Heh 1999).The global analysis is conducted by using the aero-servo-elastic code HAWC2 (Nejad et al. 2016; HAWC2 2019).

While designing offshore wind turbines, measurement-Correlation-Prediction (MCP) is a common method for estimating wind loads. Parameters needed include the wind speed, wave height, and current with a 50-year recurrence period, as required by design standard IEC 61400-3 (IEC 2009). For wind farms exposed to tropical cyclone risk, the Monte Carlo approach is used to model wind speed by a standard pressure field model, plus a tropical cyclone specific vertical wind profile model and a CFD based site-specific local terrain modification (Willis et al. 2018; Chen, Zhao, and Donelan 2013). The precise assessment of design loads for wind turbine is a complex engineering problem, requiring a coupled physics model of the air and sea and their interface to estimate loads on offshore wind turbines.

The geometry and stiffness properties of gearbox elements, including housing, shafts, bearings, and gears, can be obtained from SimPack™ software (SIMPACK 2019). Through the analyses, it is possible to predict the internal gear and bearing load reactions, and internal displacements. The critical load cases for gearbox design can thus be identified and the gearbox responses to specific load cases can be predicted.

However, deficiencies in load prediction still exist. For example, some extreme, rare-event or critical load cases are not considered in the design load spectrum. The accidental bearing and gear responses due to misalignment and bearing slip is hard to predict. The impact load leading to misalignment and gradual failure of gearbox components is difficult to assess. The relationship between complex wind loads and dynamic response of gearboxes, as well as the mechanism of component failure such as fatigue, wear and fracture are still not quite clear.

3.2. Site-Specific Design

Wind is uncertain in nature, varies from locations and weather conditions. The wind climate in mountainous complex terrain, offshore, flat terrain, and desert can be substantially different. Therefore, even the same types of wind turbines operating at different locations would have a range of operational life and different failure rate.

For example, the long-time dry and dusty low-temperature climate in onshore wind farms in northern China would cause wear damage to the blade surface and even brittle fracture. Besides, the increased viscosity of lubricant oil due to low temperature would deteriorate operating condition (Wang et al. 2013). In offshore wind turbines, failure occurs even earlier because of highly variable loads and corrosive or humid environment. Therefore, when wind turbines designed according to the wind conditions in a specific area, such as the warm temperate climate in Western Europe, are used in a cold dry climate area, such as in the north of China, the failure rate could be high.

Researchers have investigated relations between failure rate and operating conditions. Carroll et al. (2016) examined the failure rate and failure modes of wind turbine installed in several parts of Europe and presented the failure occurred during year, operational year, and failure genre of wind turbine components. Tavner et al. (2006) demonstrated that a significant cross-correlation exists between the failure rate and weather conditions. Tavner et al. (2013) confirmed that weather and location are factors closely related to wind turbine failures.

The fluctuating nature of wind speed is a considerable characteristic of location and crucial aspect of efficiency of wind turbine (Hulio 2019). The site-specific wind parameters decide the loads applied to the wind turbine blades. The increased wind loads can be the cause of decreased wind turbine performance (Cooney et al. 2017; Ganjefar and Mohammadi 2016), as well as the cause of failure of components (Nejad, Gao, and Moan 2014). The site characteristic design loads affect design assumptions and provide a cost reduction potential by site-specific design (Thomsen, Fuglsang, and Schepers 2001).

Fuglsang et al. (2001) incorporated site characteristics into the design process of wind turbines to enable site-specific design. Turbines at six different sites comprising normal flat terrain, offshore and complex terrain wind farms were investigated. The authors suggested that multiple-site wind turbines be designed for a high mean wind speed sites of flat terrain; while site-specific wind turbines be designed for low-mean-wind-speed sites and complex terrain. Hulio (2019) assessed wind characteristics at three specific sites representing the wind class 3 and 7. The site specific wind characteristics and their correlation showed that wind farm climate has significant influence on energy generation and life of components.

Therefore, the design of wind turbine gearboxes should base on the wind farm climate of the specific site. The site-specific design considers wind load characteristics under the local field operation. Proper design criteria should be selected according to the site-specific failure modes.

3.3. Structure Design

Current wind turbine gearbox designs have converged to an almost similar architecture with only a few exceptions (Musial, Butterfield, and McNiff 2007). Gearboxes in wind turbines usually use a planetary configuration that have a large speed reduction in a small volume. They usually have multiple planet gears revolving around a sun gear. Each planetary gear stages usually have three- or four-planet gears. When the number of planet gears is greater than three, flexible planet pins are used to ensure equal load sharing among planetary gears (Fox and Jallat 2008). Spur gears are selected for the low-speed stage, as they perform better at low speeds and high torques and cost less; while helical gears are chosen for the high-speed stage. The outer ring gear or annulus is connected to the rotor hub while the sun gear is connect to the generator (Ragheb and Ragheb 2010). The gear ratios among the stages are determined by a weight optimization approach (Nejad et al. 2016).

Wind turbine manufacturers, gear and bearing manufacturers, consultants, and lubrication engineers have been working together over the past two decades and established internationally recognized wind turbine gearbox design standards and codes, e.g., ISO 81400-4 (ISO 2005), or IEC 61400-4 (IEC 2012). The gearbox components, such as gears and bearings, are designed following related international standards or codes, i.e., gears by ISO 6336-2, ISO 6336-3, and ISO 6336-6 (ISO 2006), and bearings types by IEC 61400-4, etc.

The design of a wind turbine gearbox seems simple at first sight. However, despite adherence to the accepted design practices, wind turbine gearboxes still have difficulty to survive their expected design lifetime of 20-25 years. Gearboxes are usually designed according to the transmitted power or equivalent torque calculated from time series-produced torque–duration data at constant speed and constant load (Fuglsang and Thomsen 2001). However, in real engineering practice, wind turbine gearboxes operate under fluctuating load condition. The above design methodology is, therefore, obviously not an ideal approach.

So far, scarce published guidance is available on designing gearboxes operating at fluctuating speeds, fatigue or extreme loads. An expedient measure is to design components to withstand fatigue damage and ultimate loads during normal operating conditions (Nejad et al. 2016). The ultimate operational loads and relevant data can be obtained by referring to the data from similar types of wind turbine gearboxes.

Although gearboxes are not components with the highest failure rates by statistical analysis, they are reported to be one of the failures that cause the longest downtime per failure (Perez et al. 2013). The complex design, general inaccessibility, and high bearing loads are suspected to create the majority problems of wind turbines (Ragheb and Ragheb 2010).

Several approaches have been proposed to solve these problems. For example, integral planet gear bearings can be used in the first planetary gear stage, where the inner bore of the planet gear acts as the outer race for the bearing (Willis et al. 2018). Combined with the feature of increased stiffness after preloading tapered cylindrical rollers, a new bearing design could significantly reduce planet gear rim deflections, provide a life improvement of more than 150% (Willis et al. 2018). Besides, magnetically levitated bearing is an option for the high reliability gearbox (Ragheb and Reghab 2010).

More measures can be adopted to alleviate loads on the gearbox components. Modification of gear tooth profile can optimize the stress distribution on gears, thus significantly improve the reliability and safety performance of the gears. Torque dampers have been developed to dissipate torsional shock loads on gearboxes and rotors (Willis et al. 2018). To reduce maintenance demands and costs, modular design is an option to ensure easy maintenance and service. Modules with high expected failure rate can be designed separately for the convenient and quick replacement to reduce downtime.

3.4. Model-Based Design

Current design practices could not result in sufficient life for gearboxes. As the rated power of wind turbines increase, the design of gearboxes that could transmit the torque generated by longer and heavier blades with desired lifetime become an even challenging task. The precise assessment of complex dynamic loads and load response in multi-megawatt wind turbine gearboxes would facilitate an improved product development. Model-based design is a promising and efficient approach for improving design quality and identifying design deficiencies that may reduce the life of gearboxes.

Aside from the minimum requirements set by design codes, site condition, assembly and manufacturing limitations influence gearbox design. Theses information can be incorporated in the models effectively. Up to now, various models ranging from materials based simulation tools to topology optimization tools have been developed to facilitate load prediction and structural design (Willis et al. 2018). These models provide a helpful understanding of system level dynamics, damage evolution, fatigue life, and reliability.

Aeroelastic calculations are used to obtain design loads from random wind loads or extreme loads during emergency braking operations. The derivation of extreme and fatigue loads involves a large number of different load cases. Detailed models and approaches for load prediction have been introduced in Section 3.1. Nevertheless, accurate prediction of the loads for different design options or for different sites remains difficult in practical design.

The gear speed ratios of each stage can be determined by the DriveSE model based on minimizing gearbox weight in wind turbines (Guo et al. 2015). Once the optimum gear speed ratios is obtained, the gearbox components can be selected and designed.

The multibody system model is a powerful tool for analyzing load and dynamic response of wind turbine gearboxes (Dong et al. 2013; Xing, Karimirad, and Moan 2014). The natural frequencies, noise, vibration, and dynamic load of wind turbine gearboxes can be obtained by dynamic modeling. The results from the load analysis can then be used for stress analysis and reliability investigation. Several multipurpose, multibody simulation codes are available, such as SimPack™ (SIMPACK 2019).

Commercially available gear design and analysis software, as well as in-house codes or public domain bearing-rating software can be used to design gears or bearings in gearboxes. Design criteria on allowable stress and deflection are inherently included for individual gearbox components. The design criteria, together with the minimum weight objective, are used for sizing gearbox components. Several design software are available, such as Romax software (Romax 2014) for gear design, etc.

Design gearbox components separately can meet some difficulties. For example, bearing manufacturers usually do not have detailed information about gearbox system loads and responses. This may cause unpredicted bearing behavior beyond the bearing mounting location such as housing deformations, which make it difficult to analyze root-cause failure of the bearing validly (Musial, Butterfield, and McNiff 2007). A system engineering design approach representing the whole structure of gearbox is an effective tool for analyzing the stiffness of mounting at system level and the corresponding responses at critical bearing locations. Nejad et al. (2016) presents detailed information of a 5MW high-speed gearbox developed for the NREL offshore 5MW baseline wind turbine. The detailed modeling parameters required to establish the dynamic model of the gearbox are provided to offer support for large offshore wind turbine development.

System level modeling and optimization allows design, analyze and optimize wind turbine gearbox components with high reliability. According to the present design data, a physical model is built first, and the response of the present physical model is predicted under the design load. The response is then returned to the numerical optimization algorithm for design data optimization under certain constraints. After several iterations, convergence will eventually reached to the optimum design (Fuglsang and Thomsen 2001). The design optimization approaches could improve reliability, refine the design process and reduce the design-loop cycle time. These approaches contribute significantly to better design practices and lead to higher gearbox reliability anticipated by wind turbine gearbox manufacturers.

The prediction of damage evolution involves considerable challenges, including the use of finite element analysis coupled with detailed forensic analysis of failed components retrieved from the field. When the relationship between the failure process, such as crack propagation, surface wear, and the stress distribution is found, the load, stress, and reliability can be improved by structural optimization.

3.5. Reliability Design

The fluctuating nature of wind decides the loads applied to the wind turbine blades are random loads. The harsh operating condition and fatigue feature implies that reliability is common issues in the design of turbine gearboxes. Considering the difficulty and cost of maintenance, the design of wind turbine gearboxes must be assigned high reliability to reduce maintenance demands, especially for MW wind turbine gearboxes, as failure rates escalate with the increase of size and rated power of wind turbines (Echavarria et al. 2008).

As introduced previously, a wind turbine gearbox consists of low-speed input shaft, bearings, multiple planet gears, a sun gear, high-speed output shaft, etc. These components forms a complex correlated and coupled system. The components may fail individually; but more frequently, the failure of one element may trigger the failure of other elements. Besides, both single failure mode and multiple failure mode may occur in an element. The analysis of single failure mode seems straightforward; however, the analysis of multiple failure mode is much more complicated (Liu 2018). Therefore, the analysis of damage evolution of components in a gearbox brings great challenges.

Bhardwaj et al. (2019) conducted a detailed Failure Mode and Effects Analysis (FMEA) and identified the main Reliability Influencing Factors (RIF) on the failure causes. The authors also illustrated a reliability prediction method to estimate the total failure rate of an offshore wind turbine gearbox. Hoseynabadi et al. (2010) used FMEA technique to improve the reliability of wind turbine system. The authors studied 2000 kW wind turbine of different configuration and compared to imaginary wind turbine system. Guo et al. (2009) examined the reliability of wind turbine by incomplete data. They utilized the three Weibull parameter distribution function including Maximum likely hood and Least Squares for life testing of wind turbine components. Echavarria et al. (2008) presented a comparative study of reliability over the time of advancement of wind turbine technology. The authors observed the significant differences in reliability for certain subcomponents, which is depend on the size and on the type of power control. Unlike the general wind turbines whose failure rates decrease with time, multi-megawatt wind turbines showed a longer early failure behavior and significantly higher annual failure rate of components per wind turbine, implying more efforts are required to improve reliability.

The FMEA and fault tree analysis (FTA) can be used in the reliability design of gearboxes. Components in a gearbox have different failure rate and different degree of accessibility for maintenance. For example, high-speed shaft have a high failure rate in a wind power gearbox. Due to the inconvenience of disassembly, the failure of high-speed shaft may cause long downtime and high maintenance cost. The design should, therefore, not only assign reliability for the elements, but also consider the convenience of disassembly. Such design could achieve more accurate and tighter design margins with high reliability. However, little open publications have been noticed reporting about it so far.

4. MANUFACTURING APPROACHES TO IMPROVE RELIABILITY OF WIND TURBINE GEARBOXES

4.1. Heat Treatments and Manufacturing

The design life of wind turbine gearboxes is expected as 15-20 years. During the service life, wind turbine gearbox materials will experience degradation, aging and deterioration, such as defect expansion and damage evolution. This process is greatly related to the original microstructure, which is determined by manufacturing and heat treatment. Proper manufacturing and heat treatment could improve material properties and produce long life and failure resistance materials, and eventually improve the wind turbine gearbox reliability.

Gear failures characterized by superficial damage and tooth breakage are common and intricate. To achieve a long lifetime, gears should not only have a tough core to withstand fatigue bending and high impact loads, but also should have a hard surface and high fatigue strength to resist wear and pitting.

The versatile mechanical properties can be achieved by employing appropriate heat treatments. Gears in wind turbine gearboxes generally use alloy steels, e.g., 40Cr, 16MnCr5, with heat treatment of normalizing, through hardening, carburizing, and nitriding. The sun gears, planets and helical gears are case-hardened carburized, while the ring gears are usually quenched and tempered. Carburizing or nitriding generates residual compressive stresses in the superficial layers. It will produce hard outer surface to resist contact fatigue, fretting and wear; while at the same time, protect the core from becoming brittle, and eventually increase the endurance limit to against cyclic stresses (Willis et al. 2018). The case-hardened gears and bearings have higher contact fatigue endurance limits, and could prevent surface initiated pitting and subsurface initiated micropitting. Laser quenching is a promising heat treatment method, as it could produce deeper hardened layer with less deformation (Xing 2009). Sometimes two or more heat treatment processes are combined to achieve the desired properties.

It is no doubt that stringent manufacturing and tolerance requirements benefit the improvement of reliability. Low surface roughness has a significant effect on preventing micro-pitting, reducing friction and wear. It will also improve lubrication conditions between tooth surfaces, and eventually prolong the service life of gears (Lin, Shen, and Zhao 2012). Although field-failure assessments indicate that only up to 10% of gearbox failures may be related to gear manufacturing quality (Musial, Butterfield, and McNiff 2007), adherence to accepted gear and bearing industry practices could definitely benefit the increase of reliability.

4.2. Assembly

Considering the size and weight of a MW turbine gearbox, precise assembly presents a challenging task. At the premise that most gearbox components have been designed and fabricated by the best industry practices available, assembly is of paramount importance in the process of wind turbine gearbox manufacturing, as it takes longer and has a great impact on the gearbox performance. Assembly error and damage is another direct root cause of the failure of wind turbine gearboxes.

When assembly gearboxes, important assembly requirements, such as eccentricity, backlash, clearance, etc., must be carefully controlled and satisfied. Under fluctuating loads, the assembly induced errors, misalignment, as well as structural deformation, have direct influence on the occurrence of most mechanical fatigue failures.

Several points require special attention during assembly, i.e., the alignments and clearances between bearing inner ring and outer ring, the alignments and clearances of meshing gears, the axial motion of sun-pinion, shafts, and bearings, etc. When elements are connected by interference fit, the external load, the amount of interference, and coefficient of friction, will affect the strength of connection (Li 2008) and must be properly controlled.

Improper assembly, including insufficient or excessive bolt preloading torque, excessive gear meshing clearance, inadequate alignment accuracy, pits on the surface of components, etc., may cause damage to the components and should be avoided. The study of the accumulation of assembly errors could improve the design and manufacture process to achieve high reliability wind turbine gearboxes.

4.3. Maintenance

Maintenance aims to prevent or postpone unforeseen events and damages by different types of repair (Echavarria et al. 2008). To achieve high reliability, availability, maintainability and safety, a complex maintenance procedure is required. However, maintenance is far more expensive, especially for offshore wind turbines, as the turbines may be difficult to access due to harsh weather conditions, such as wind and waves. Besides, when failures occur, the day rate for offshore crane from the offshore oil and gas industry is at least 10 times higher than a land based crane for similar lifting heights (Van Bussel and Zaaijer 2001).

Preventive maintenance has the potential to reduce considerably the operational and maintenance costs. A comprehensive design should include recommendations and comments on the maintenance planning and procedures. Nejad et al. (2016) developed a detailed inspection and maintenance planning map based on the fatigue damage of gears and bearings. During routine inspection and maintenance, the maintenance team can focus on those components with the highest probability of failure rather than by examining all gears and bearings, thus reduce time and efforts to identify the source of problem.

Effective maintenance strategies can minimize failure loss by detecting and repairing defects at early stage. Real-time monitoring systems are used to monitor the gearbox, main shaft and generator bearings in real time (Willis et al. 2018). Noise, vibration, high oil temperature, abnormal oil pressure are the early signs of failure. When abnormal signals appears, intensive observation and detection of gearbox is required to find out the location of the fault in time, and deal with it properly.

The gearbox monitoring technologies commonly rely on accelerometer-based vibration and thermocouple-based oil temperature measurements. Accelerometers are usually used to measure vibration to identify pitting or wear on gears or bearings. Temperature monitoring is arranged to measure temperature at the inner rings, the outer rings, and planet bearings. Lubrication monitoring includes oil temperature, lubricant cleanliness, and oil debris observation (Musial, Butterfield, and McNiff 2007; Salameh et al. 2018). Cooling system are important for reducing temperature and filtration system can remove all particles larger than seven microns in size. Both measures could improve oil lubrication and ensure turbines more reliable. Magnetic particle filters is another option to filter out generated wear debris down to ~5 mm in size (Needelman, Barris, and LaVallee 2009).

Conclusion

As the global wind power capacity increases, the development of large rated power, long service, high reliability wind turbines has become trends in the next 10 years. However, current gearboxes have yet to achieve their design life of twenty years. The higher than-expected failure rates require identifying root causes of failures first and then developing mitigation measures that may advance the design and manufacturing of wind turbine gearboxes.

This paper first presents common failure modes of wind turbine gearboxes and analyzes their causes. Principal deficiencies in design and primary inadequacies in manufacturing that contribute to substantial losses of operational life for most wind turbine gearboxes are then addressed. Measures to improve the reliability and prolong the operational lifetime of MW rated wind turbine gearboxes through innovative design and manufacturing approaches are finally proposed. Hopefully, by advanced design, manufacturing and on-line monitoring, high reliability gearboxes can be developed to operate confidently under complex loading conditions.

REFERENCES

Arabian-Hoseynabadi, Hooman, Hashem Oraee, and Peter Tavner. 2010. "Failure Modes and Effects Analysis (FMEA) for wind turbines." *International Journal of Electrical Power & Energy Systems* 32 (7): 817-824.

Bhardwaj, U., A. P. Teixeira, and Carlos Guedes Soares. 2019. "Reliability Prediction of an Offshore Wind Turbine Gearbox." *Renewable Energy* 141: 693-706.

Carroll, James, Alasdair McDonald, and David McMillan. 2016. "Failure Rate, Repair Time and Unscheduled O&M Cost Analysis of Offshore Wind Turbines." *Wind Energy* 19 (6): 1107-1119.

Chen, Shuyi S., Wei Zhao, and Mark A. Donelan. 2013. "Directional Wind-Wave Coupling in Fully Coupled Atmosphere-Wave-Ocean Models: Results from CBLAST-Hurricane." *Journal of the Atmospheric Sciences* 70 (10): 3198-3215.

Cooney, Ciaran, Raymond Byrne, William Lyons, and Fergal O'Rourke. 2017. "Performance Characterisation of a Commercial-Scale Wind

Turbine Operating In an Urban Environment, Using Real Data." *Energy for Sustainable Development* 36: 44-54.

Dong, Wenbin, Yihan Xing, Torgeir Moan, and Zhen Gao. 2013. "Time Domain-Based Gear Contact Fatigue Analysis of a Wind Turbine Drivetrain under Dynamic Conditions." *International Journal of Fatigue* 48: 133-146.

Echavarria, E., Berthold Hahn, Gerard van Bussel, and Tetsuo Tomiyama. 2008. "Reliability of Wind Turbine Technology through Time." *Journal of Solar Energy Engineering-Transactions of the ASME* 130 (3): 031005.

Fox, Gerald, and Eric Jallat. 2008. Use of the Integrated Flexpin Bearing for Improving the Performance of Epicyclic Gear Systems. *Timken Technical Report*.

Fuglsang, Peter, and Kenneth Thomsen. 2001. "Site-Specific Design Optimization of 1.5-2.0 MW Wind Turbines." *Journal of Solar Energy Engineering-Transactions of the ASME* 123 (4): 296-303.

Ganjefar, Soheil, and Ali Mohammadi. 2016. "Variable Speed Wind Turbines with Maximum Power Extraction Using Singular Perturbation Theory." *Energy* 106: 510-519.

Greco, Aaron, Shuangwen Sheng, Jonathan Allen Keller, and Ali Erdemir. 2013. "Material Wear and Fatigue in Wind Turbine Systems." *Wear* 302 (1-2): 1583-1591.

Guo, Haitao, Simon Watson, Peter Tavner, and Jiangping Xiang. 2009. "Reliability Analysis for Wind Turbines with Incomplete Failure Data Collected from after the Date of Initial Installation." *Reliability Engineering & System Safety* 94 (6): 1057-1063.

Guo, Y., T. Parsons, R. King, K. Dykes, and Paul S. Veers. 2015. *An Analytical Formulation for Sizing and Estimating the Dimensions and Weight of Wind Turbine Hub and Drivetrain Components*. Technical Report NREL/TP-63008, National Renewable Energy Laboratory, Golden, Colorado, USA.

Hawc2. 2019. "*Aero-Servo-Elastic Calculation Code for Horizontal Axis Wind Turbine*." Accessed July 24. http://www.hawc2.dk.

Hulio, Zahid Hussain. 2019. "*An Assessment of Wind Characteristics, Performance Characterization and Availability of Site Specific Wind Farms.*" PhD diss., Dalian University of Technology.

IEC. 2009. *Wind turbines - Part 3: Design requirements for offshore wind turbines.* IEC 61400-3-2009. Switzerland: International Electrotechnical Commission.

IEC. 2012. *Wind Turbines - Part 4: Design Requirements for Wind Turbine Gearboxes.* IEC 61400-4-2012. Switzerland: International Electrotechnical Commission.

Igba, Joel, Kazem Alemzadeh, Christopher Durugbo, and Egill Thor Eiriksson. 2017. "Through-life Engineering Services of Wind Turbines." *CIRP Journal of Manufacturing Science and Technology* 17: 60-70.

Igba, Joel, Kazem Alemzadeh, Christopher Durugbo, and Keld Henningsen. 2015. "Performance Assessment of Wind Turbine Gearboxes Using In-service Data: Current Approaches and Future Trends." *Renewable & Sustainable Energy Reviews* 50: 144-159.

ISO. 2005. *Wind Turbines – Part 4: Design and Specification of Gearboxes.* ISO 81400-4:2005. Switzerland: International Organization for Standards.

ISO. 2006. *Calculation of Load Capacity of Spur and Helical Gears. Part 2: Calculation of Surface Durability (Pittings).* ISO 6336-2:2006. Switzerland: International Organization for Standards.

ISO. 2006. *Calculation of Load Capacity of Spur and Helical Gears. Part 3: Calculation of Tooth Bending Strength.* ISO 6336-3:2006. Switzerland: International Organization for Standards.

ISO. 2006. *Calculation of Load Capacity of Spur and Helical Gears. Part 6: Calculation of Service Life under Variable Load.* ISO 6336-6:2006. Switzerland: International Organization for Standards.

Jantara, V. L. Junior, H. Basoalto, H. Dong, F. P. G. Marquez, and M. Papaelias. 2018. "Evaluating the Challenges Associated with the Long-term Reliable Operation of Industrial Wind Turbine Gearboxes." *International Conference on Materials Engineering and Science* 454: 1-8.

Li, Ming. 2008. "*Research on Longitudinal Correction of Helical Gear in Gearbox for Wind Turbine Gearboxes.*" MA thesis, Dalian University of Technology.

Lin, Tengjiao, Liang Shen, and Junyu Zhao. 2012. "Fatigue Life Finite Element Analysis of Output Gear Pair of Wind Turbine Speed-Increase Gearbox." *Journal of Chongqing University* 35 (1): 1-6.

Lin, Yonggang, Le Tu, Hongwei Liu, and Wei Li. 2016. "Fault Analysis of Wind Turbines in China." *Renewable & Sustainable Energy Reviews* 55: 482-490.

Liu, Huahan. 2018. "*Mechanical Reliability Models with Failure Dependency and its Application.*" PhD diss., Dalian University of Technology.

Mott, Robert L. 2003. *Machine Elements in Mechanical Design*. New Jersey: Prentice Hall.

Musial, Walt, Sandy Butterfield, and Brian McNiff. 2007. "Improving Wind Turbine Gearbox Reliability." *European Wind Energy Conference and Exhibition 2007*, Milan, Italy, May 7-10.

Needelman, William M., Marty A. Barris, and Gregory L. LaVallee. 2009. "Contamination Control for Wind Turbine Gearboxes." *Power Engineering* 113 (11): 112-120.

Nejad, Amir R., Zhen Gao, and Torgeir Moan. 2014. "On Long-term Fatigue Damage and Reliability Analysis of Gears under Wind Loads in Offshore Wind Turbine Drivetrains." *International Journal of Fatigue* 61: 116-128.

Nejad, Amir R., Yi Guo, Zhen Gao, and Torgeir Moan. 2016. "Development of a 5 MW Reference Gearbox for Offshore Wind Turbines." *Wind Energy* 19 (6): 1089-1106.

Nikas, George K. 2010. "A State-of-the-Art Review on the Effects of Particulate Contamination and Related Topics in Machine-Element Contacts." *Proceedings of the Institution of Mechanical Engineers Part J-Journal of Engineering Tribology* 224: 453-479.

Perez, Jesús María Pinar, Fausto Pedro García Marquez, Andrew Tobias, and Mayorkinos Papaelias. 2013. "Wind Turbine Reliability Analysis." *Renewable & Sustainable Energy Reviews* 23: 463-472.

Ragheb, Adam M., and Magdi Ragheb. 2010. "Wind Power Gearbox Technologies." *Proceedings of the 1st international Nuclear and Renewable energy conference*, Amman, Jordan, March 21-24.

Romax. 2019. "Romaxwind software." Accessed July 24. https://romaxtech.com/software/.

Sørensen, Poul, A. K. Unnikrishnan, and Sajan A. Mathew. 2001. "Wind Farms Connected to Weak Grids in India." *Wind Energy* 4 (3): 137-149.

Salameh, Jack P., Sebastien Cauet, Erik Etien, Anas Sakout, and Laurent Rambault. 2018. "Gearbox Condition Monitoring in Wind Turbines: a Review." *Mechanical Systems and Signal Processing* 111: 251-264.

SIMPACK. 2019. "*Multi Body System Software.*" Accessed July 24. http://www.simpack.de.

Tavner, Peter J., Clare Edwards, Andy Brinkman, and Fabio Spinato. 2006. "Influence of Wind Speed on Wind Turbine Reliability." *Wind Engineering* 30 (1): 55-72.

Tavner, Peter J., David Michael Greenwood, M. W. G. Whittle, Rosa Gindele, S. Faulstich, and Berthold Hahn. 2013. "Study of Weather and Location Effects on Wind Turbine Failure Rates." *Wind Energy* 16 (2): 175-187.

Thomsen, Kenneth, Peter Fuglsang, and G. Schepers. 2001. "Potentials for Site-Specific Design of MW Sized Wind Turbines." *Journal of Solar Energy Engineering-Transactions of the ASME* 123 (4): 304-309.

Toft, Henrik Stensgaard, lasse Svenningsen, Wolfgang Moser, John D. Sørensen, and Morten Lybech Thøgersen. 2016. "Wind Climate Parameters for Wind Turbine Fatigue Load Assessment." *Journal of Solar Energy Engineering-Transactions of the ASME* 138 (3):031010.

Van Bussel, Gerard, and Michiel Zaaijer. 2001. "Reliability, Availability and Maintenance Aspects of Large-Scale Offshore Wind Farms, a Concepts Study." *Proceedings of Marec*.

Veers, Paul S., and Steven R. Winterstein. 1998. "Application of Measured Loads to Wind Turbine Fatigue and Reliability Analysis." *Journal of Solar Energy Engineering-Transactions of the ASME* 120 (4): 233-239.

Wang, Hui, Xiaolong Li, Gang Wang, Dong Xiang, and Yiming Rong. 2013. "Research on Failure of Wind Turbine Gearbox and Recent

Development of Its Design and Manufacturing Technologies." *China Mechanical Engineering* 24 (11): 1542-1549.

Willis, D. J., C. Niezrecki, D. Kuchma, E. Hines, S. R. Arwade, R. J. Barthelmie, M. DiPaola, Patrick J. Drane, Christopher J. Hansen, Murat Inalpolat, John Hunter Mack, A. T. Myers, and Mario Rotea. 2018. "Wind Energy Research: State-of-the-Art and Future Research Directions." *Renewable Energy* 125: 133-154.

Wilson, R., Walker, S. and Heh, P. 1999. *Technical and User's Manual for the FAST_AD Advanced Dynamics Code.* Oregon State University, Corvallis, Oregon.

Xing, Dazhi. 2009. "Strengthen Methods of the Ring Gear in Wind Turbine Gearbox." *Machinist Metal Forming* 17: 14.

Xing, Yihan, Madjid Karimirad, and Torgeir Moan. 2014. "Modelling and Analysis of Floating Spar-Type Wind Turbine Drivetrain." *Wind Energy* 17 (4): 565-587.

INDEX

A

action fields, 52
advantages, 6, 9, 62, 88, 112, 128
ambitious targets, 11
artificial bee colony, 77, 130
assessment, viii, 2, 9, 28, 30, 31, 32, 37, 40, 47, 50, 67, 175, 179
authorities, 7, 21, 22, 26, 27
awareness, vii, ix, 3, 17, 36

B

barriers, ix, 12, 31, 33, 36
benefits, 9, 12, 22, 60, 62, 63, 64, 70
biomass, viii, 2, 6, 7, 10, 12, 15, 18, 20, 22, 25, 81, 85, 86, 88, 90, 93, 100, 101, 103, 106, 109, 126, 133, 137, 149, 150, 151, 152, 155, 163

C

carbon, 3, 11, 13, 22, 28, 33, 36, 49
challenges, ix, 3, 5, 20, 31, 35, 36, 54, 147, 173, 181, 182, 189

circular economy, 16, 17
cities, viii, 2, 4, 5, 6, 9, 10, 11, 12, 13, 15, 17, 18, 19, 20, 21, 24, 26, 29, 32, 62
climate, vii, ix, 1, 2, 4, 11, 15, 16, 26, 35, 36, 37, 40, 47, 48, 49, 50, 52, 57, 60, 62, 63, 69, 71, 78, 174, 176, 177
climate change, viii, 1, 2, 4, 11, 15, 26, 36, 37, 40, 50, 52, 57, 62, 63, 69, 71, 78
climate change issues, 40
climate change mitigation, 11, 63
combined heat and power, 10, 15
communities, 13, 17, 31, 59, 67
competition, 20, 22, 31, 59, 61, 64, 65, 73
competitiveness, ix, 12, 29, 36, 64
concept, 4, 5, 9, 16, 29, 31, 109
conclusion, 24, 26, 45, 46, 53, 56, 69, 77, 147, 186
consolidation, 15, 27
consumers, 10, 16, 19, 24, 27, 29
consumption, vii, 3, 5, 8, 10, 13, 16, 24, 58
cooperation, ix, 7, 13, 36, 38, 39, 48, 49, 55, 69
cost, x, xii, 5, 7, 12, 22, 31, 165, 167, 177, 178, 182, 183

D

decision makers, 7, 37, 45, 46, 60
deficiencies, 176, 179, 187
deformation, 172, 173, 184, 185
deployment, 4, 7, 26, 28, 30, 36, 49, 58, 68, 150
deregulated market, 23
deregulation, 23, 29
design and manufacturing, vii, xii, 166, 167, 168, 187
designing, 36, 77, 90, 91, 125, 126, 147, 175, 178
disadvantages, 6, 88, 112
disruptive innovation, 55, 65, 67
distribution, vii, ix, x, 5, 16, 30, 51, 53, 56, 57, 58, 69, 70, 174, 179, 181, 182
district cooling system, 15, 27, 32
district heating, viii, 2, 5, 14, 15, 24, 25, 27, 29, 30, 31, 32, 33
diversification, 3, 10, 15, 16, 19, 23, 27, 39, 45

E

ecological threats, 3
economic factors and stakeholder awareness, vii, ix, 36
economic growth, 32, 52
economic problem, 16
economic progress, viii, 2
economic research, 4, 6, 37
economic theory, 16
ecosystems, 8
eco-village, 17
electricity, vii, ix, 15, 23, 27, 29, 31, 51, 52, 53, 55, 56, 57, 58, 60, 61, 62, 63, 65, 66, 67, 70, 71, 72, 73
electricity generation, vii, ix, 51, 53, 55, 56, 57, 58, 63, 68, 70, 79, 80
electricity transition, 52, 55, 56, 57, 58

emissions, vii, 1, 2, 7, 11, 24, 27
energy, vii, viii, ix, x, xi, xii, 1, 2, 3, 4, 5, 6, 7, 8, 9, 10, 11, 12, 13, 14, 15, 16, 17, 19, 20, 21, 22, 23, 24, 25, 26, 27, 28, 29, 30, 31, 32, 33, 35, 36, 37, 39, 45, 46, 47, 48, 50, 52, 53, 55, 56, 57, 58, 59, 60, 61, 62, 63, 64, 65, 66, 67, 68, 69, 71, 73, 74, 165, 166, 167, 177, 191
energy consumption, 3
energy demand, x, 4, 7, 12, 76, 82
energy efficiency, 12, 23, 27, 31, 46, 48
energy independence, 4
energy policy, 4, 11, 20, 22, 24, 26, 30, 36, 64, 83
energy poverty, 4
energy prices, viii, 2, 15
energy sector, ix, 5, 11, 26, 36, 37
energy security, 3, 15
energy self-sufficiency, 4
energy supply, viii, 1, 2, 4, 5, 15, 17, 23, 26, 27, 63, 68
energy system(s), vii, ix, x, 4, 7, 12, 20, 27, 30, 33, 35, 36, 37, 39, 46, 53, 76, 82, 85, 86, 87, 88, 90, 100, 104, 114, 125, 138, 147, 150, 157, 158, 159, 160, 161, 162
engineering, 10, 47, 50, 175, 178, 181
environment(s), 3, 12, 17, 26, 28, 40, 52, 54, 57, 58, 64, 69, 167, 169, 170, 176
environmental issues, 9
environmental protection, viii, 1, 4
environmental threats, vii, 1

F

failure modes, vii, xii, 166, 168, 169, 170, 172, 173, 176, 177, 187
financial, 5, 12, 19, 46, 62
financial resources, 5
financial support, 62
forests, 8
formation, 10, 13, 17, 40, 60, 61

fossil fuels, viii, 2, 3, 23, 56, 58, 63, 78, 84, 86
functions, 5, 26, 109
funding, 15, 19, 26
funds, 13, 17, 22, 27
fuzzy logic, 40, 50
fuzzy PROMETHEE, 36, 37, 40, 45

G

geographical database, 21
geothermal, 6, 7, 8, 13, 15, 18, 25, 32, 85, 86
governance, 32, 52
growth, xii, 3, 61, 62, 65, 68, 165, 166

H

heat pumps, 12
heat sector, 6, 7, 12
housing, 168, 171, 172, 173, 175, 181

I

incentives, 9, 61
incumbents, 55, 59, 63, 64, 65, 67, 68, 69, 70
industry, 3, 10, 17, 23, 45, 47, 59, 65, 66, 184, 185
inexhaustibility, 9
inexhaustible, 7
infrastructure, viii, 2, 5, 6, 9, 10, 14, 15, 16, 20, 22, 27, 66
integrated approach, 12, 32
integration, viii, 2, 3, 5, 6, 9, 10, 11, 23, 27, 29, 39, 48, 68, 111, 157
intermittent nature, 5
issues, 4, 6, 11, 13, 16, 64, 65, 68, 69, 182

K

key points, 19
knowledge, v, viii, x, 1, 2, 3, 4, 5, 6, 9, 10, 13, 18, 27, 28, 37, 52, 76, 78, 112, 132, 147, 148, 149
knowledge structure, 6
knowledge-based economy, viii, 2, 4, 5, 28

L

laws, 4, 15, 17, 27, 65
legitimacy, 52, 53, 56, 57, 58, 61, 62
legitimations, ix, 51, 53, 61
lifetime, 65, 168, 175, 178, 179, 183, 187
local communities, 13
local government, 5, 12, 26
low-carbon technologies, 36

M

manufacturing, vii, xii, 166, 167, 168, 170, 172, 180, 183, 184, 185, 187
market maturity, vii, ix, 36
methodological principle, 9
methodology, 178
monopoly, 10, 16, 19, 22, 31, 58, 59
multicriteria decision aid, vii, ix, 36, 37
multicriteria decision aid methods, 37
multidimensional, 40
multidisciplinary, 3
municipalities, 2, 5, 9, 13, 14, 20, 22, 23, 26, 27

N

natural monopoly, 10, 16, 19, 22, 31, 58
negative consequences, viii, 2, 27, 70
niches, 54, 58, 62

O

oil, 21, 22, 23, 24, 25, 37, 169, 171, 176, 185, 186
opportunities, ix, 5, 9, 19, 26, 27, 37, 47, 49, 51, 58
optimization, x, xi, 59, 178, 180, 181

P

pathway(s), vii, ix, 7, 35, 38, 39, 40, 45, 46, 48
plants, 12, 23, 24, 29, 30, 31, 63, 67, 69
policy, ix, 4, 10, 11, 20, 21, 22, 23, 24, 26, 28, 30, 35, 36, 37, 38, 40, 45, 47, 48, 49, 62, 64, 65, 72
policy initiative, 23
political power, 58
politics, 56, 58
politics of layering, 61
pollution, viii, xii, 2, 3, 15, 26, 62, 165, 166, 169
potential, 5, 7, 10, 13, 15, 19, 32, 33, 88, 111, 117, 149, 152, 177, 186
power generation, vii, x, xii, 66, 165, 166, 167
power plants, 10, 58, 59, 63, 66, 69
practical implementation, 5, 14, 17
pricing, 11, 15, 16, 20, 22, 23, 31, 33
principles, 14, 17, 53, 54, 64
private investment, 6
private sector, 13, 32
problems, viii, 1, 3, 4, 5, 6, 9, 10, 12, 14, 16, 17, 26, 37, 52, 57, 58, 64, 65, 70, 77, 113, 168, 172, 179
producers, 12, 16, 20, 24, 27, 59, 63, 65
profit, 11, 19, 22, 23, 24
program, 15, 29, 160
progress, viii, 2, 4, 5, 7, 10, 11, 14, 19, 150, 169
public benefit, 6

R

regulation(s), 11, 16, 20, 21, 31, 33, 68, 73, 148
regulatory changes, 61, 65, 70
regulatory framework, vii, ix, 36, 40, 54, 69
reliability, vi, vii, xi, xii, 4, 11, 13, 76, 97, 104, 165, 166, 167, 168, 174, 179, 180, 181, 182, 183, 184, 185, 186, 187, 188, 190, 191
renewable energies, 52, 72, 73
renewable energy, vii, viii, 1, 2, 3, 4, 6, 7, 12, 13, 15, 16, 25, 26, 27, 28, 29, 30, 32, 33, 46, 47, 49, 53, 61, 62, 63, 85
renewable energy sources (RES), viii, xi, 1, 2, 3, 4, 5, 6, 7, 8, 9, 10, 11, 13, 14, 15, 16, 17, 18, 19, 22, 23, 24, 25, 26, 27, 28, 29, 30, 32, 47, 48, 49, 77, 80, 83, 85, 90, 99, 151, 152
renewable energy technologies, 6, 13, 15, 26
renovation, 13, 16, 33
requirement(s), vii, x, 21, 22, 68, 166, 180, 184, 185, 189
RES directive, 22
RES technologies, viii, 2, 3, 5, 6, 7, 9, 10, 11, 27
resources, x, 3, 5, 6, 9, 11, 13, 16, 27, 69
response, 175, 176, 179, 180, 181
restructuring, 23
review, vi, 13, 30, 31, 32, 109, 157, 162, 165, 166, 190, 191
role of municipalities, 14, 21
root(s), 57, 168, 169, 170, 171, 181, 185, 187

S

safety, 46, 167, 168, 179, 185
security, viii, 2, 4, 15, 27, 62, 63, 70
society, 12, 13, 27, 60, 69, 70

Index 197

software, 12, 77, 88, 89, 91, 94, 102, 103, 110, 111, 112, 113, 175, 180, 191
solar, x, 3, 6, 7, 8, 9, 12, 13, 15, 18, 25, 28, 60, 66, 68, 69, 72, 76, 82, 85, 86, 91, 92, 93, 94, 97, 101, 102, 103, 110, 111, 115, 116, 117, 118, 119, 121, 122, 123, 124, 126, 127, 128, 133, 136, 138, 140, 141, 146, 147, 149, 150, 151, 152, 157, 158, 159, 160, 162, 163, 188, 191
state, 5, 7, 10, 19, 20, 21, 24, 26, 56, 66, 167
strategic planning, 20, 21, 29, 32
stress, 169, 170, 172, 179, 180, 181
structure, 6, 19, 44, 73, 171, 181
support, 3, 11, 16, 17, 20, 21, 22, 23, 24, 25, 26, 33, 35, 40, 46, 47, 57, 60, 62, 63, 65, 66, 67, 68, 181
sustainability, 11, 12, 26, 29, 32, 47, 50
sustainable development, 4, 8, 9, 10, 11, 12, 14, 17, 22, 26, 29, 32, 188
sustainable energy, vii, ix, 3, 5, 14, 26, 27, 30, 36, 38

T

technological innovation, 7
technologies, viii, ix, x, 1, 2, 3, 4, 5, 6, 7, 9, 10, 11, 13, 16, 19, 20, 24, 25, 27, 36, 37, 62, 63, 186
technology, vii, viii, ix, 2, 8, 20, 25, 28, 35, 36, 48, 60, 62, 63, 65, 68, 73, 182
temperature, 10, 13, 169, 171, 174, 176, 186
territorial, 4, 5, 9, 11, 13, 14, 16, 26, 29
territorial aspect, 4, 9, 11
trade deficit, 3
transformation, v, 2, 17, 19, 28, 35, 38, 39, 46, 51, 53, 55, 56, 65, 67, 72

transformation of energy infrastructure of municipalities, 2
transformation pathways, 38, 39, 46
transition, ix, 11, 13, 20, 21, 22, 24, 31, 33, 36, 37, 52, 53, 55, 56, 60, 66, 67, 69, 73, 74

W

waste, 8, 15, 16, 23, 25, 27
wind, vi, vii, x, xii, 3, 6, 18, 20, 25, 53, 60, 63, 66, 69, 73, 76, 77, 81, 82, 85, 86, 88, 89, 90, 91, 92, 93, 94, 95, 97, 100, 101, 103, 104, 106, 108, 110, 113, 125, 126, 133, 136, 138, 139, 141, 147, 148, 149, 150, 151, 152, 157, 158, 159, 160, 161, 162, 163, 165, 166, 167, 168, 169, 170, 171, 172, 173, 174, 175, 176, 177, 178, 179, 180, 181, 182, 183, 184, 185, 186, 187, 188, 189, 190, 191, 192
wind farm, 175, 176, 177
wind power, 166, 183, 186
wind speed, 170, 174, 175, 177
wind turbine gearbox, vii, xii, 165, 166, 167, 168, 169, 170, 172, 173, 174, 175, 177, 178, 179, 180, 181, 182, 183, 184, 185, 187
wind turbines, xii, 60, 165, 166, 167, 169, 170, 175, 176, 177, 179, 180, 182, 183, 185, 186, 187, 189

Z

zero waste technologies, 16
zoning of DH areas, 21

Related Nova Publications

Bioeconomic and Policy Aspects of Future Sustainable Biofuel Production

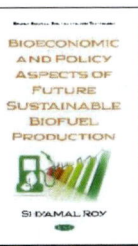

Author: Shyamal Roy

Series: Energy Science, Engineering and Technology

Book Description: This book states developments in the bioenergy market and related policies. Recent bioenergy developments, often induced by policies, lead to a greater connection between energy and agricultural markets and influenced relative food and feed prices and land-use changes.

Softcover ISBN: 978-1-53616-136-6
Retail Price: $82

Can Biofuels Alleviate the Energy and Environmental Crisis?

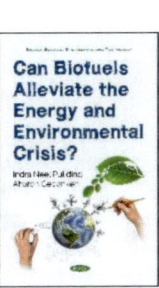

Authors: Indra Neel Pulidindi and Aharon Gedanken

Series: Energy Science, Engineering and Technology

Book Description: The objective of the compilation of the book titled "Can Biofuels Alleviate The Energy & Environmental Crisis?" is to reach out to policy makers, scientists, industrialists, and students with a message as well as scientific strategies for alleviating the twin problems of energy and environmental crisis posing a threat to future generations.

Hardcover ISBN: 978-1-53615-050-6
Retail Price: $195

To see a complete list of Nova publications, please visit our website at www.novapublishers.com

Related Nova Publications

SPACETIME ENERGY: A TWENTY FIRST CENTURY PERSPECTIVE

EDITOR: B. G. Sidharth

SERIES: Energy Science, Engineering and Technology

BOOK DESCRIPTION: The anthology titled "Spacetime Energy: A Twenty First Century Perspective" encompasses novel and avant garde ideas that delineate the intrinsic nature of spacetime and several of its aspects.

HARDCOVER ISBN: 978-1-53615-585-3
RETAIL PRICE: $160

SOLAR ENERGY SYSTEMS: PROGRESS AND FUTURE DIRECTIONS

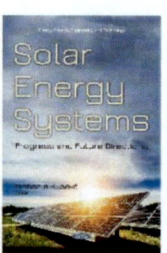

EDITOR: Michaela Rybová

SERIES: Energy Science, Engineering and Technology

BOOK DESCRIPTION: *Solar Energy Systems: Progress and Future Directions* presents some new concepts and ideas regarding future steps in the development and progress of solar thermal energy. Preliminary results for advanced control of solar plants are presented using effective defocusing mechanisms.

SOFTCOVER ISBN: 978-1-53616-142-7
RETAIL PRICE: $82

To see a complete list of Nova publications, please visit our website at www.novapublishers.com